EXPLORATIONS IN CHEMISTRY

A MANUAL FOR DISCOVERY

NICHOLAS KILDAHL
Worcester Polytechnic Institute

THERESA VARCO-SHEA
Wellesley College

John Wiley & Sons, Inc.
New York • Chichester • Brisbane • Toronto • Singapore

ISBN 0-471-12699-3

Printed in the United States of America

10 9 8 7 6 5 4 3 2 1

Preface

Features of the Manual

Explorations in Chemistry: A Manual for Discovery is the result of our efforts to create a laboratory program that responds to widely recognized needs in chemical education. **First, all of the experiments in the manual are designed in the discovery format**. Students are expected to carry out laboratory work to verify or refute a hypothesis developed at the outset of the laboratory period. Student data are pooled and the hypothesis is examined in light of the collective experimental results. In the course of this process, important principles of chemistry are discovered from student data. The experience is intended to convey, to at least some extent, the manner in which science is actually done. A traditional laboratory approach, in which an experiment is performed in a "cookbook" manner to verify a previously-discussed principle, conveys none of the excitement of experimental science. The element of discovery drives scientific investigation. Without it, science is a rote and seemingly dull process in which we demonstrate what we already know. Of the two varieties of discovery--"guided" and "open"--we have opted for a guided approach, to avoid problems of time and materials that often arise in an open discovery format. **Second, a substantial number of the experiments involve the use of spectroscopic and chromatographic instrumentation**. Again, the purpose is to give students a sense of how modern chemistry is done. Laboratory is more interesting and exciting for students when they answer questions using the same means that practicing chemists use. Excitement and interest are further enhanced if the instrumentation is computer driven. **Third, with a couple of exceptions, experiments are designed to require only small quantities of reagents**. The cost and safety benefits of this feature are obvious. In addition, it provides the instructor with the opportunity to point out that modern instrumentation can provide a lot of information about a very small sample of material. **Fourth, many of the experiments are designed to be carried out by teams of students, working in cooperative fashion**. This is intended to take advantage of the many benefits that derive from cooperative learning activities, as well as to introduce the students to the project team approach that is prevalent in business and industrial settings. **Fifth, four experiments are presented in a project format, designed to encourage students to work in a research style setting**. Depending on the desires of the instructor, students could be asked to report the results of their projects in scientific journal format, or to present oral reports of their results. In our experience, reports written by the students are of good quality, and improve with each successive project. Oral reports presented at the end of the project, **Beverages**, are enjoyed by all and are quite informative. The project experiments may be modified and adapted to suit the instructor. **Finally, the use of computers for data collection, manipulation, and presentation, and the use of spreadsheet programs for data handling are emphasized in this manual**. This is an important aspect of the modern chemistry laboratory that should, if possible, be introduced in the laboratory portion of education.

Flexibility in Design

Flexibility is a strong feature of Explorations in Chemistry: A Manual for Discovery. Experiments are not numbered, and no particular order need be followed. The manual is thus suitable to a variety of laboratory programs. Although many experiments require the use of instrumentation, many do not. In addition, all of the experiments involving electronic absorption spectroscopy may be performed using Spectronic 20 spectrometers. The manual is therefore suitable for schools of varying size and resource availability. The manual provides tear-out **Results** pages for laboratory programs which do not require students to keep notebooks. However, for programs which require notebooks, detailed information is presented for keeping a good laboratory notebook. The **Background** section of each experiment closes with one or more questions to be used as a springboard for hypothesis development. Generally, however, the experiments may be used to answer any of a number of questions which may be of more interest to an instructor than those that we propose. This gives an element of flexibility to the very important hypothesis development phase of each experiment. Many experiments are designed to allow students to learn from mistakes, and to repeat operations more than once to improve technique. More important, perhaps, is that time is allowed for the students to think about what they are doing and trying to accomplish. We have attempted to include many different types of experiments so that the student receives a broad exposure to many aspects of chemistry. Therefore, there are experiments that emphasize techniques, such as **Volumetric Measurements and Solutions, Titrations, Synthesis of $Ni(NH_3)_6Cl_2$**, and **Synthesis of Hydrated Copper(II) Sulfate**. However, we want to teach more than techniques with these experiments so the students are asked to go beyond simple "cookbook" chemistry. The prelab **Preparation** exercises for the synthesis of $Ni(NH_3)_6Cl_2$ and $CuSO_4{\bullet}5H_2O$ require the student to think about the percent yield of the reaction and then to scale the synthesis to prepare the required amount of product.

Organization

The organization of the manual requires some comment. **Several introductory chapters deal with special topics which are important background for the students.** These include safety considerations; experimental error; weighing procedures; the use of volumetric glassware; the analysis of titration data; and the keeping of a laboratory notebook. **Following the introductory chapters are the laboratory experiments.** These are not numbered, and may be taken in any order according to the needs of the instructor. The first 11 experiments require no instrumentation beyond the analytical balance, and may be performed with common laboratory equipment. Following these are 14 instrument-based experiments, involving the use of gas chromatography or spectroscopy (FTIR, UV-Vis, and NMR). **Next, five workshop-type activities are included**, on topics ranging from graphical analysis of data to computer spreadsheets and molecular models. The workshops may be used as chemical-free laboratory exercises or as homework assignments, as fits the instructor's needs. **Finally, four projects, each requiring a minimum of 2 weeks, are presented at the end of the manual.** Each laboratory experiment consists of the following sections:

- An opening statement indicating the duration of the experiment and whether it is to be performed individually or in teams.
- The **Goals** section lists suggested learning objectives for the experiment.
- The **Background** section, which provides introductory information about the experiment. This section should be read and understood by students before they come to the laboratory.
- The **Equipment and Materials** required for the experiment are listed next. Each school has a system for making materials available and the instructor will explain the procedure for the institution.
- Immediately before the experimental section for each experiment is a note on **Safety** for that experiment. Students should understand the safety precautions before beginning any experiment. In addition, they should understand how to use all of the equipment and instrumentation.
- The **Experimental** section gives directions on how to proceed with the experiment. Sometimes, the directions are very specific and detailed. In other cases, students must devise their own procedures with minimum guidance from the manual. For these experiments, students are asked to complete many of the necessary calculations in the **Preparation** (see below).
- **Disposal Methods** are provided after the **Experimental** section. Since each state has different regulations, these are only guidelines and the specific directions given by the laboratory instructor will take precedence. Proper waste disposal is an important part of good lab practice and should be taken seriously.

•The **Questions** following the **Disposal Methods** are answered after the experiment is complete. Some questions may be answered based on a student's own data but others may require consideration of the class data as a whole. The questions are to promote thought about the experiment and the data. They will help students draw conclusions about the data.
•Several highlighted questions at the end of the **Background** section are the basis for the lab discussions.
•Immediately following the experiment pages is a tear-out **Preparation** page. The **Preparation** includes reading assignments in the manual and general suggestions for reading in the course text. In addition, several sample problems are given to provide preparation in the type of calculations involved in the experiment. The instructor must provide guidelines for when the **Preparation** is due. Most of the experiments are difficult to complete without proper preparation.
•The last section for an experiment is the tear-out **Results** page. This provides an alternative to the laboratory notebook for recording of data.

We have developed or tried a number of teaching techniques in order to maximize the effectiveness of the manual. We discuss the most successful of these here.

Group Work. A number of experiments, in particular those involving instrumentation and the projects, work best when done by teams of several students. We have used groups of up to 5 students, with 3 and 4-member teams being common. Despite some dissatisfaction initially, students usually enjoy the group work and participation. They often express surprise that they can learn so much from each other.

In some cases we have found it useful to model group structure after that used in the chemical industry. Thus a Project Director may be appointed from within a student group. His or her responsibilities are 1) to direct the research program, including setting goals and deadlines; 2) to consult with the instructor, either orally or via a written proposal, about the group's program; 3) to finalize the form and content of all reports, including research proposals, written reports, and oral presentations, to which all members of a group must contribute. The remaining members of the group serve as Bench Chemists. They are responsible for their own experiments and also for contributing to the writing and editing of reports and proposals.

The organization of the groups emphasizes several key points: 1) The importance of keeping good notes in the lab notebook. The success of the project depends on the results of each contributor and requires that each person record accurate and detailed procedures and observations. 2) The importance of writing and speaking skills. The successful scientist (or professional) must communicate effectively both orally and in writing. 3) The challenges and rewards of working as part of a team.

Self-Evaluation. For the group projects, the students evaluate their own work and that of the other group members. These evaluations are used to determine the group participation part of the lab grade. Students are usually honest and frank on the evaluations. In addition, evaluations are usually very consistent within a group.

Notebook Exchange. For institutions which require lab notebooks, it is a good exercise at some point during the term or semester to have the students exchange notebooks at the halfway point of a lab period to see whether they can duplicate another student's results using only the procedures written in the notebook. This serves to drive home the point that sufficient experimental detail must be written in the notebook that the experiment can be repeated by reference to the notebook alone. While students tend to become frustrated by the exercise, they get the point. The experiments, **Nuts and Bolts**, **Synthesis of $Ni(NH_3)_6Cl_2$**, and **Synthesis of Hydrated Copper(II) Sulfate**, work well for this exercise.

Hypothesis Development. The **Background** section of each experiment closes with questions which may be used as a guide for hypothesis development. We generally suggest several possible answers to each question, then ask the students to vote their preference. The winning answer is adopted as the hypothesis to be tested during the laboratory period.

Data-Based Discussion. In conducting the discussion of the students' results, particularly as they relate to the hypothesis developed at the outset of the laboratory period, it is best if possible to use the data actually obtained by the students during the lab period, rather than data obtained by an earlier group of students. To facilitate this, we ask students to enter data to a computer file as they obtain it. The data can then be quickly plotted and the plot converted to an overhead transparency for use in discussion. The value of using the students' own data is that they see, first, the generally very good quality of the data; and second, whether or not their own data fits in well with that of the class as a whole.

Instructors teaching a laboratory course from this manual should expect to be challenged. When students are asked to think for themselves, many new and interesting questions can be expected. The instructor should expect to hear fewer questions of the "how do I...?" type and more of the "why does it work that way?" type. We are very pleased by the excitement this type of laboratory work generates in our students.

Acknowledgments

Finally, we would like to acknowledge the efforts of our colleagues in connection with Explorations in Chemistry: A Manual for Discovery. TV-S would like to acknowledge and to thank M. Turnbull for dedicated teaching of the revised labs at Wellesley; B. Shaw of the University of Connecticut for helpful suggestions which served as guidelines for the group format adopted; the faculty and staff of Wellesley college for their support and contributions to this work; N. Doe for unending faith and lab work during the summer of 93; C. Haddleton and H. Clark for providing stockroom support during all of the revisions; J. Hann for permission to reproduce parts of her notebook; and J. Darlington, dedicated teacher, friend and scientist, for her unfaltering belief in and support of this work. NKK would like to acknowledge and to thank Ladislav Berka for his many contributions to the laboratory experiments, his collaboration in making the lab program at WPI a success, and his general moral support; Doug Burns for his computer expertise, his creativity, and his excellent teaching in the lab; Jim Pavlik, for his unflagging support of the laboratory development effort; and Dave Messier and Jack Ferraro, for technical support.

Table of Contents

Introduction

Chemistry is a laboratory science. Every law and theory of chemistry is founded on facts discovered by systematic and careful effort in the laboratory. You are about to embark on what may be your first laboratory experience. It is the purpose of Explorations in Chemistry: A Manual for Discovery to make that experience both interesting and informative for you.

We have several objectives in this manual:

●To provide experiments with an element of **discovery**. By this we mean that you will discover laws and principles of chemistry from your experimental results. This is the way real chemistry happens. The discovery approach incorporates 4 facets:

- ■Formulation of a question to be answered;
- ■Development of a hypothesis (a reasonable answer to the question);
- ■Exploration of the hypothesis through experiment;
- ■Modification of the hypothesis or experiment (if necessary) to answer the question.

We do not expect you to formulate original questions, develop hypotheses, or design experimental procedures on your own. The experiments have been written to help you with each stage. What we do expect is that by doing the experiments in discovery mode, you will gain an appreciation for the manner in which science is actually done.

●To provide a **modern** laboratory experience. Modern chemistry relies on data obtained from a wide array of sophisticated instrumentation. We intend to give you exposure to hands-on use of instrumentation as part of this program. Toward that end, a number of experiments in this manual are **instrument-based**; that is, the data required in hypothesis testing are acquired instrumentally.

●To emphasize the connection between the macroscopic viewpoint of the laboratory and the molecular view of the chemist. Throughout the laboratory course the focus is: what do the experiments tell us about what the molecules are doing, and how is what they are doing related to their structures?

●To provide experience in teamwork and in group dynamics. This experience should prove to be invaluable preparation for your eventual work experience, where team approaches to problem solving are becoming the norm.

●To provide experience in laboratory applications of computers.

●To provide a laboratory experience which is enjoyable, academically challenging, and relevant.

Your instructor will decide which of the experiments in the manual you are to do, and will inform you sufficiently in advance for you to prepare yourself for each experiment. **There will not necessarily be, nor should you expect, perfect coordination between subject material in lecture and lab**. After all, our purpose in lab is to enable you to discover some principles of chemistry. That would be much less interesting if you had already encountered the principle in lecture! The minimum preparation required for an experiment is to 1) Read the entire experiment; and 2) Complete the **Preparation** page(s) for the experiment before coming to lab. Most of the experiments are difficult to complete properly without preparation. Once in the lab, you will work individually, with a partner, or in groups depending on the experiment and the materials and equipment available. If you are inadequately prepared, you will put yourself and your teammates at a disadvantage. Please take the responsibility to understand the experiment as well as you possibly can in advance.

Organization of the Manual

Each experiment is organized into several sections. These are listed below:

•**Goals**. These are statements of how we expect you to benefit from each experiment.

•**Background**. This section may be short or quite extensive. It provides necessary information about the experiment. This section should always be read and understood before the lab. You should ask any questions concerning the **Background** or **Preparation** well before coming to lab. Your lab instructor's office hours are a good time to sort out any problems. At the end of the **Background** section are one or more highlighted questions. It is the goal of the experiment to answer the question(s) via hypothesis development and experimentation.

•**Equipment and Materials**. The equipment and materials available are listed next. Occasionally, all of the materials may not be available so substitutions will be made. Each school has a system for making these materials available and your instructor will explain the procedure for your institution.

•**Safety**. It is particularly important to be aware of any safety hazards involved in the experiment.

•**Experimental**. This section gives directions on how to proceed with the experiment. Sometimes, the directions are very specific and detailed. However, often you must devise your own procedures with minimum guidance from the manual. Many of the necessary calculations must be completed in the **Preparation**.

•**Disposal Methods**. Disposal guidelines are provided after the **Experimental** section. Since each state has different regulations, these are only guidelines and the specific directions given by the laboratory instructor will take precedence. Proper waste disposal is an important part of good lab practice and should be taken seriously.

•**Questions**. These are to be answered following completion of the experiment. The answers constitute part of the laboratory report. Some questions may be answered based on your own data but others may require you to consider the class data as a whole. The questions are to help you think about the experiment and the data, and to draw conclusions about the data.

•**Preparation** page. Immediately following the experiment pages is the **Preparation** page. The preparation includes reading assignments both in the manual and in your text. You will also find words or concepts to define and practice problems. Your instructor will tell you when the **Preparation** is due.

•**Results** page. This is for reporting results. Whether you use this page or a laboratory notebook will be decided by your instructor.

The manual also contains several reference sections that deal with special topics important to your work. Some covered topics include weighing procedures, errors in measurements, the use of volumetric glassware, the analysis of titration data, and the keeping of a laboratory notbook. The **Preparation** directs you to read the appropriate sections of the manual before beginning the experiment. These sections are intended to be a general reference for you throughout the course.

Four projects at the end of the manual provide experience in the team approach to problem solving. The manual emphasizes the team approach because real life situations in industry, medicine, business, and research more often than not require that you work with groups of people. A major goal of this course is to help you to function effectively and comfortably in this type of group working environment. Within each group, the instructor may assign a project director and bench chemist(s). These positions may rotate through the term or semester so that each student is a project director at least once. All students in a group carry out lab work, contribute to the experimental design and reports, and are responsible for understanding all techniques and chemistry performed by other group members. In addition, these projects may involve a system of peer review. In other words, you will be asked to grade your own work as well as that of your group members. Your instructor will indicate when and if a peer review is to be used.

Laboratory Attendance.

Laboratory attendance is mandatory, and you are expected to attend the lab to which you have been assigned unless your lab instructor gives you permission to do otherwise. Every effort will be made to help students make up labs missed due to illness or serious personal emergencies. Please follow these procedures if you miss a laboratory:

1) Notify the lab instructor by telephone or e-mail when it is clear that lab will be missed.
2) As soon as possible, discuss the reason for missing the lab with the lab instructor. If appropriate, the lab instructor will then suggest a time for completing the missed lab.
3) It is then the student's responsibility to arrange to complete the lab.

Laboratory Regulations and Safety Procedures

Strict safety regulations will be enforced in the chemistry laboratory. The rules are listed below, with first-aid procedures for common lab mishaps. Notify the instructor if you are pregnant or think you may be pregnant. Special, confidential arrangements will be made to allow you to work safely in the lab.

Eye protection
•Safety goggles must be worn <u>at all times</u>.
•Contact lenses are not permitted.

Personal protection
•Wear proper clothing--no shorts, halter tops, sandals, or cloth shoes (shoes should shed liquid).
•Long hair must be tied back.
•No smoking, drinking, eating, or gum chewing.
•Never work alone.
•No horseplay or unauthorized experiments.
•Know the location and operation of all safety equipment: fire extinguishers, showers, eye washes, fire blankets.
•When inserting glass into a stopper, moisten both glass and stopper with water first. Protecting your hands with a towel, grip the glass about an inch above the stopper and push with a slight rotation. (Failure to observe this procedure leads to many lab accidents. Please be careful.)
•NEVER smell a chemical by putting your nose into a container.
•Wash with soap and water before leaving lab.

Chemical waste
•Dispose of all waste materials as directed.
•Use the proper waste containers, never mix wastes.

Using heat
•Never point the open end of a test tube toward anyone.
•Check for flammables **before** lighting a Bunsen burner.
•Never leave a flame or heat source unattended.
•Never heat a closed system.

Handling chemicals
•Reactions which produce objectionable odors must be run in the hood.
•DO NOT PIPET BY MOUTH. Use a rubber bulb or pipet pump for suction.
•NEVER pour water into concentrated acid. Pour the acid slowly into the water with constant stirring.
•Reagent bottles:

  a. Read the labels carefully.
  b. Do not contaminate reagent bottles.
    1. Transfer what you need into a beaker.
    2. Pour with the label facing up.
    3. Do not insert anything into a reagent bottle.

4. Never return unused chemicals to reagent bottles.

c. Do not take more reagent than is required for an experiment.

Other procedures

•Report all injuries to the lab instructor or stockroom personnel.

•After an experiment, turn off gas and water; clean up your desk and the area around your balance.

•Performing unauthorized experiments will result in course failure. If you want to carry out an experiment to test an idea, clear it with your instructor first.

•Always read the safety section for each experiment.

First Aid

Call for help immediately if you suffer an injury or a spill!

For any spilled acids or bases, first aid should begin by washing with lots of water. **ALWAYS USE WATER FIRST**.

1. For base burns, follow the water wash by rinsing with a 5% ammonium chloride solution. Then wash again with water.
2. For acid burns, follow the water wash by rinsing with a sodium bicarbonate solution; then wash with water.

DO NOT APPLY OINTMENTS OR SALVES TO A SERIOUS BURN EXCEPT UNDER THE DIRECTION OF A PHYSICIAN. CONTACT YOUR INSTRUCTOR FOR ASSISTANCE.

Uncertainty, Precision, and Error in Experimental Results

Measurement of quantities such as length, volume, mass, temperature, light absorbance, and time is the heart of experimental science. Ultimately, the most elegant theory rests on numbers obtained in the laboratory by a scientist using a measuring device or instrument. No device or instrument is capable of producing an exact result--a result with no uncertainty. When a scientist reports the result of a single measurement, such as a length or volume, s/he also reports the uncertainty in the result, a number indicating the reliability of the measuring device used. For example, the mass of a sample obtained using a balance which weighs to the nearest 0.1 g may be 32.0 g. The uncertainty in this number is 1 unit in the tenths decimal place, and indicates that the balance reading is uncertain by 1 unit in this place. The mass is reported as 32.0 ± 0.1 g, and the result of the weighing is known to 1 part in 320. Scientists sometimes communicate uncertainty implicitly rather than explicitly. For example, the mass above might be reported as 32.0 g. Reporting the tenths decimal place implies an uncertainty of 1 digit in this place. The mass is known to 3 significant figures, which is the number of digits written (except in numbers like 0.0032, where leading zeros are place holders, not significant figures). Weighing this sample on a more sensitive balance might give the result 32.0492 g. Now the uncertainty is 1 in the ten-thousandths place, and the mass is known to 6 significant figures. The result of this weighing is known to 1 part in about 320000, much better than with the less sensitive balance.

Conversion Between Scales

Most quantities can be expressed in more than one system of units. For example, length can be expressed in English units or metric units; temperature can be expressed in Celsius degrees, Fahrenheit degrees, or Kelvins. Generally, uncertainties are related by the same conversion factors as relate the quantities in two scales. If a length measured in inches has an uncertainty of 1 inch, then the same length expressed in centimeters (cm) has an uncertainty of 2.54 cm, because 1 in = 2.54 cm. If a temperature on the Celsius scale has an uncertainty of 0.1 °C, the same temperature on the Fahrenheit scale will have an uncertainty of 0.18 °F because there are 1.8 Fahrenheit degrees per Celsius degree. Some examples will make this clear.

<u>Example.</u> A man is 74 inches tall. What is his height in meters?

Height = 74 inches, with an implicit uncertainty of ± 1 inch. There are 2.54 cm per inch. Height in cm is

$$74 \times 2.54 = 187.96 \text{ cm}$$

The uncertainty in this number is ±2.54 cm. Since the uncertainty occurs first in the 1's place, we report the height as 188 ± 3 cm. Height in meters is 1.88 ± 0.03. The usual significant figure rule would force us to report this as 1.9 m, because the height in inches is given to 2 significant figures. But this would imply that we know the man's height to only 1 part in 19 in meter units, whereas we know it to 1 part in 74 in inches. Reporting the height as 1.9 m would mean a loss of information!

<u>Example.</u> What is 9 °C on the Fahrenheit scale?

The relationship between the two scales is

$$°C/100 = (°F - 32)/180$$

Substituting 9 for °C and solving for °F gives 48.2 °F. How many significant figures should we report? The significant figure rule (see your textbook) tells us to report 50 °F, with 1 significant figure. But according to this rule, 7 °C would be 40 °F, while 8, 9 and 10 °C would all be 50 °F! Clearly this significant figure rule is not working well. The key here, as above, is to realize that 1 Celsius degree is equivalent to 1.8 Fahrenheit degrees. An uncertainty of 1 degree in Celsius temperature will translate to an uncertainty of 1.8 degrees in Fahrenheit temperature. Since uncertainty occurs first in the 1's place, the correct Fahrenheit temperature above is 48 °.

Reproducibility and Accuracy

The precision in a result indicates the reproducibility of the result--how well we can expect to reproduce it when we make repeated measurements. The higher the precision, the better the reproducibility. Precision is different from uncertainty, which has meaning only for a single measurement, as explained above.

Example. 3 drivers, A, B, and C, make the trip along Route 9 from Spags in Shrewsbury to Pizzeria Uno in Natick. From his initial and final odometer readings, driver A reports the distance as 22.4 miles. Drivers B and C report the distance as 23.6 and 21.0 miles, respectively. Find a) the uncertainty in driver A's measurement; b) the best estimate of distance from Spags to Uno, based on the efforts of these three drivers; c) the precision (reproducibility) of this best estimate.

a) The uncertainty in driver A's result is a measure of how well he can read his odometer. Most odometers read to 0.1 mile, meaning we can estimate a single odometer reading to ± 0.05 mile. A distance, the difference between two odometer readings, is probably good to only 0.1 mile. The uncertainty in driver A's distance is ± 0.1 mile; he should report his result as 22.4 ± 0.1 mile. Similarly, drivers B and C should report their results as 23.6 ± 0.1 miles and 21.0 ± 0.1 miles.

b) The best estimate of the distance is the average or mean of the results of the three drivers.

$$\text{Mean} = (22.4 + 23.6 + 21.0)/3 = 22.3 \text{ miles}$$

c) The precision of this number is a measure of how close any driver could expect to come to 22.3 miles, driving his own car. The precision may be estimated as 1/2 the difference between the maximum and minimum distances obtained by our 3 drivers, because the result obtained by any particular driver would probably fall within this range.

$$\text{Precision} = (23.6 - 21.0)/2 = 1.3$$

The best estimate of the distance from Spags to Uno from our data is 22 ± 1 miles. We write it this way, rather than as 22.3 ± 1.3, because the precision tells us that distances measured by various drivers will reproduce to no better than 1.3 miles. There is no point in reporting the average to any better than the 1's place! Occasionally, 1/2 the range of a series of measurements will be smaller than the uncertainty in a single measurement. In these cases, one would use the uncertainty in a single measurement rather than 1/2 the range to represent the precision, as reproducibility cannot be better than the uncertainty in a particular measurement.

A more sophisticated way to report the precision in a measurement is to use the standard deviation. If a series of n measurements of a quantity, x, are x_1, x_2, ..., x_n and the mean of the values is x_{av}, then the standard deviation, σ_x, in x is calculated as follows:

$$\sigma_x = [\Sigma\,(x_{av} - x_n)^2/(n-1)]^{1/2}$$

This formula says to 1) calculate the deviation of each individual measurement from the mean and square it; 2) add up all the squared deviations; 3) divide the sum by 1 less than the number of measurements; and

4) take the square root of the result. The standard deviation in our Spags to Uno distance is

$$\sigma = \{[(-0.1)^2 + (-1.3)^2 + (1.3)^2]/(2)\}^{1/2} = 1.3$$

Note that in this example, this is the same as half the range. Although 1/2 the range and standard deviation will generally be similar, they will not in general be equal. The standard deviation represents a statistically more significant measure of precision than does 1/2 the range.

The error in a measured result is the difference between the measured value and the true value of the quantity being measured. Error indicates the accuracy of the measured value. Error is to accuracy what precision is to reproducibility. An extension of the example above shows the meanings of these terms clearly.

Example. An accurate road survey gives the distance between Spags and Uno as 23.9 miles. What is the error in our estimate of this distance?

Error = measured value - true value = 22.3 - 23.9 = -1.6 miles. The negative sign indicates that the measured value is smaller than the true value.
In summary:

Precision--the reproducibility of a measurement, indicated by writing the measured value ± range/2 or ± S.D.;
Accuracy--the difference between the measured value of a quantity and the true value. Indicated by writing the error (measured value - true value).

Uncertainties in Calculated Results

The uncertainty in a measured number carries into calculations made using the number. A calculated result is expressed consistently with the uncertainties in the numbers used in the calculation. How this is done depends on the arithmetic operation performed.

Addition and Subtraction. The uncertainty in the result of adding or subtracting two numbers is the sum of their uncertainties.

Suppose that a buret is filled with solution to a volume reading of 2.64 mL. Solution is delivered from the buret to a flask. The final buret reading is 27.48 mL. What volume of solution was delivered? The uncertainty in a single buret reading is ± 0.01 mL. The volume delivered is the difference between final and initial readings, and has an uncertainty which is the sum of the uncertainties in the two readings:

Volume delivered to flask = 24.84 ± 0.02 mL.

Multiplication and Division. The relative uncertainty in a product or quotient is the same as that of the least-well-known factor. The relative uncertainty is expressed using the "x parts in y" approach introduced above. If a number known to 1 part in 100 is multiplied by a number known to 1 part in 1000, the product can be known to at best 1 part in 100.

For example, suppose a 10.00 mL sample of water (delivered from a 10-mL volumetric pipet) has a mass of 9.9643 g (measured with an analytical balance). What is its density? Since density is mass per unit volume, we calculate

$$\rho = 9.9643 \text{ g}/10.00 \text{ mL} = 0.99643 \text{ g/mL}$$

This must be reported to a number of significant figures consistent with the uncertainties in mass and volume:

Uncertainty in mass is 0.0002 g, or 2 parts in 99643.

Uncertainty in volume is 0.01 mL, or 1 part in 1000.

The volume is less certain than the mass. <u>The density cannot be known more precisely than the least-precisely-known of mass and volume.</u> At best, density is known to about 1 part in 1000. We should report density as 0.996 ± 0.001 g/mL. (The usual significant figure rule would allow us to report density to 4 significant figures, as 0.9964 g/mL, since volume is known to 4 sig figs. However this implies that we know density to 1 part in 9964! How can this be if we only know volume to 1 part in 1000??) A common sense approach to significant figures using the "x parts in y" method avoids the pitfalls of the textbook rules.

Precisions carry through calculations according to the same rules.

Logarithms. Frequently we need to take the logarithm of a measured number to make a plot. How do we express the uncertainty in the logarithm of a measured value?

Suppose we need the logarithm of $(1.25 \pm 0.01) \times 10^{-4}$. We first find the log of 1.25×10^{-4} to be -3.9031. To find the uncertainty in this, we compute logs of the maximum and minimum values our measured value could have, based on its uncertainty:

$$\log (1.26 \times 10^{-4}) = -3.900$$
$$\log (1.24 \times 10^{-4}) = -3.907$$

Comparing with 3.9031, the variation is in the 3rd decimal place. Uncertainty in the log is 1/2 of 0.007, or 0.004. The result is 3.903 ± 0.004. The number of digits following the decimal point in the log is the same as the number of significant figures in the original measured value. This is always true.

Exponentiation and Trigonometric Functions. The same approach used for logarithms can be used when raising a number to a power or finding the values of trigonometric functions of a number:

1. Apply the function to the measured value.
2. Find the extremes of the measured value by first adding, then subtracting the uncertainty.
3. Apply the function to the extremes to get the extremes of the function.
4. Find the range of values of the function as the difference between its extremes.
5. The uncertainty in the function is 1/2 the range.

For example, raise 10 to the power -2.4 ± 0.2.

1. $10^{-2.4} = 3.981 \times 10^{-3}$
2. Extremes of measured value are -2.2 and -2.6.
3. $10^{-2.2} = 6.310 \times 10^{-3}$; $10^{-2.6} = 2.512 \times 10^{-3}$
4. Range of function $= 6.3\text{-}2.5 \times 10^{-3} = 3.8 \times 10^{-3}$
5. Uncertainty in function $= 1.9 \times 10^{-3}$.

The value of the function is reported $(4 \pm 2) \times 10^{-3}$. Note how an uncertainty of 1 part in 12 in the original value translates to uncertainty of 2 parts in 4 in the exponential. Exponential functions magnify uncertainty!

Reading Graduated Measuring Devices

Many laboratory measurement devices are graduated--marked with equally spaced lines corresponding to incremental values of the quantity measured. For example, a meter stick is 1 meter long, with large lines for centimeter subdivisions, and smaller lines for millimeter subdivisions. It is graduated in millimeters. A 100-mL graduated cylinder is marked with large lines every 10 mL and smaller lines every mL. It is graduated in mL. Similarly, burets are graduated in 0.1 mL, 10-mL measuring pipets to 0.1 mL, and standard laboratory thermometers to 1 °C.

There is a standard procedure for reading a value from a graduated device, summarized as follows: estimate the value to one decimal place more than the level of graduation. When reading a temperature with a thermometer graduated in degrees, estimate to tenths of a degree. Record a buret reading to the nearest hundredth of a mL (if the liquid level falls exactly on a major graduation, say at 32.00 mL, write 32.00, not 32). Record a volume measured with a 100-mL graduate to the nearest 0.1 mL. Please adhere strictly to this procedure. If you do not, you will introduce into your measurement more uncertainty than actually exists.

References

1."Error, Precision, and Uncertainty" by Charles J. Guare. *Journal of Chemical Education*, **1991**, 68, 649-652.
2. "Is 8 °C Equal to 50 °F?" by H. Bradford Thompson. *Journal of Chemical Education*, **1991**, 68, 400-402.
3. "Basic Principles of Scale Reading" by Gavin D. Peckham. *Journal of Chemical Education*, **1994**, 71, 423-424.

Determining Mass

The balance used to determine the mass of chemicals or equipment used in lab depends on the number of significant figures required. The maximum mass that the balance can accommodate must also be considered. For approximate estimates of mass, a double or triple beam balance is used. These balances are good to $\pm$ 0.01 g. If more accurate masses are required an analytical balance is used. All balances must be kept clean. The use of the balance will be demonstrated.

General Guidelines for Using Balances

1. Never place chemicals directly onto the pan.
2. Weigh by difference to minimize errors.
3. Clean up after using the balance.
4. Return all settings to zero when done.

Guidelines for Using the Analytical Balance

1. Weighing paper should not be used since the transfer from the paper will eliminate the accuracy achieved.

2. Addition or removal of liquid or solid from a container must be carried out outside the balance case.

3. All objects must be at room temperature. Hot objects will cause convection currents in the balance. Cold objects will tend to condense moisture.

4. Objects that have been heated to a high temperature and then cooled in a desiccator may be prone to pick up moisture from the air when first put on the balance pan. All solid surfaces tend to hold a trace of moisture to a degree that is determined by the humidity of the air and, unless it is harmful to the experiment, a more reliable and reproducible result will be obtained if water-vapor equilibrium is established between the surface and the air before recording the mass. Constancy of mass with time is a good indication that equilibrium has been established. Objects that have recently been very wet and wiped dry may actually lose mass as water vapor leaves the surface and equilibrium is established.

5. Hygroscopic or volatile materials should always be kept in closed containers and the mass found as rapidly as possible.

6. Your skin contains oils and moisture which could be transferred to the weighing bottle. To avoid contamination use clean tongs, or a band of lintless paper to handle the weighing bottle.

General Procedures for Finding the Mass of a Solid -- Mass by Difference

There are two proper procedures:

1. If the sample need not be transferred quantitatively from the weighing vessel, find and record the mass of a watch glass or beaker. Add the sample. Record the mass of sample and container and calculate the mass of the sample.

2. For hygroscopic materials or for substances which must be transferred quantitatively to another vessel (e.g. a volumetric flask), an excess of the sample is put into a clean dry weighing bottle. Find the mass of the bottle with sample, transfer the sample, and find the mass of the bottle with sample again. The mass of the sample is calculated by difference.

Following is an example of the type of data to be recorded for this procedure:

Initial bottle plus sample:	20.4319 g
Final bottle plus sample:	18.6799 g
Mass of sample:	1.7520 g

The Mass of a Liquid

Liquids must always be contained in a tightly stoppered flask or weighing bottle when in or on the balance.

General Rules of Analytical Balance Use

The analytical balance is a highly accurate, sensitive, and delicate instrument. Great care is required in its use.

1. Do not move a balance, or shock it by dropping something on the pan. Place objects on the pan gently.

2. Do not place chemicals directly on the balance pan. Place the object or substance in a beaker, weighing bottle, or a watch glass.

3. Keep the balance and balance area clean. If you spill a chemical inside the balance compartment, carefully brush the material out of the balance, then wipe it up.

4. Do not attempt to make any adjustments to a balance. Notify the instructor if adjustment is needed.

Use Of Volumetric Glassware

There are several laboratory devices used to measure liquid volumes. The one chosen in a particular situation depends on the precision required and the purpose of the measurement.

- Graduated cylinder--used when high precision is not needed;
- Volumetric flask (TC)--used to prepare solutions of precisely known concentration;
- Buret (TD)--used to deliver a variable but precise volume of liquid to a receiver;
- Transfer pipet (TD)--used to transfer a specific precise volume from one container to another;
- Measuring pipet (TD)--used to transfer a range of specific precise volumes.

Devices marked TC are calibrated to contain a certain volume of liquid. To contain means that when filled to the calibration mark, the device contains exactly the calibration volume (e.g., 500.00 mL). Devices marked TD are calibrated to deliver (TD) precise volumes (e.g., 10.00 mL). For a buret or measuring pipet, this means that the volume of liquid delivered (that is, drained) from the device is given precisely by the difference between the final and initial liquid levels in the device. For a transfer pipet, to deliver means that when first filled to the calibration mark, then allowed to drain as completely as possible, the volume delivered (the volume of liquid that drains out) is exactly the calibration volume (e.g., 10.00 mL) of the pipet. To deliver pipets must be held vertically for 30 seconds after the liquid apparently has finished draining and the last drop of solution dislodged by touching the tip to the side of the receiver. There is inevitably liquid left inside a to deliver device after draining is complete, meaning that it actually contains more than the calibration volume when full. This residual liquid should be left in the device, not ejected into the receiving vessel. Table 1 at the end of this section lists some common glassware and the uncertainty in volume measured with each.

Approximate Volume Measurements

Graduated beakers and flasks may be used to make very rough volume measurements with an uncertainty on the order of 10%. Graduated cylinders (graduates) are used to make volume measurements with an uncertainty on the order of 1%. The quantity of liquid in a graduate is read by matching the bottom of the liquid meniscus (curved surface at the top of the liquid column) with the graduation marks on the side of the graduate. If the liquid is opaque or forms a convex meniscus, read the position of the top of the meniscus. Your line of sight should be perpendicular to the scale to avoid false readings due to parallax error (Figure 1). Graduates come in a variety of sizes ranging from 5 mL to 2000 mL. A common size is 100 mL, graduated to 1 mL. The uncertainty in a single reading of a 100 mL graduate is 0.5 mL.

Figure 1
Parallax Error

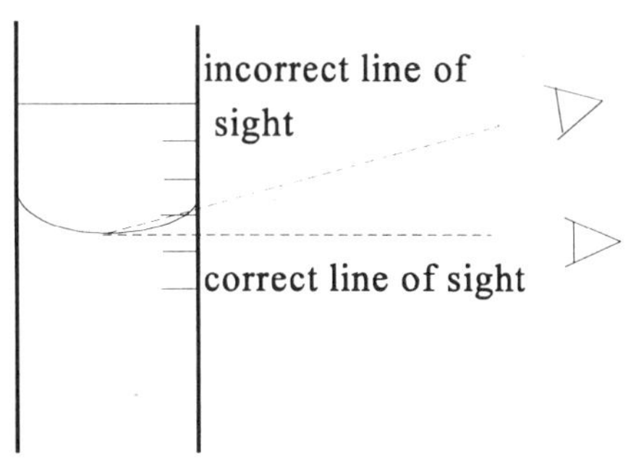

<u>Precise Volume Measurements</u>

Precise volume measurements are made with volumetric glassware: pipets, burets, and volumetric flasks. All volumetric glassware must be scrupulously clean to give accurate results. Cleanliness is indicated if water spreads evenly over the inside surface of the glass without forming any droplets. Droplets indicate the presence of grease, dirt, or some other contaminant which must be removed before use. For assistance in obtaining clean volumetric glassware, see your instructor. Your instructor will demonstrate the proper use of volumetric glassware.

Buret

Burets are used for delivering variable but precisely known volumes of liquid. Common laboratory burets have total capacities of 25 or 50 mL, and are graduated to 0.1 mL. A buret reading can therefore be estimated to the nearest 0.01 mL. The uncertainty in a single buret reading is usually taken to be 0.01 mL. Because a delivered volume is the difference between two buret readings, the uncertainty in delivered volume is 0.02 mL.

A. **Operating Procedure**. Put the buret in a buret clamp with the stopcock handle to the right. Circle the buret with your left hand and operate the stopcock with your thumb and first two fingers.

B. **Checking for Cleanliness**. Fill the buret with distilled water to above the 0.00 mL mark, then let water run out the tip into a spare beaker. Examine the inner wall of the buret between the 0.00 and 50.00 mL marks for water drops. No droplets means the buret is clean and you may rinse the buret with reagent.

Drops on the inside wall means the buret is dirty. It should be cleaned with cleaning solution or exchanged for a clean one. Proceed only with a clean buret.

C. **Rinsing with Reagent Solution**. Place ≈ 30 mL reagent solution in a clean, dry beaker. Empty the buret and close the stopcock. Pour ≈ 5 mL reagent solution into the buret and run ≈2 mL out the tip into a spare beaker. Take the buret out of the clamp and rotate it slowly while pouring the rest of the reagent out the top. Rinse two more times.

D. **Titrating Samples**. Titration is the process of controlled addition of a substance to a second substance with which it reacts. A solution of the first substance, the titrant, is contained in a buret. The second substance, also in solution, is kept in an Erlenmeyer flask. The Erlenmeyer flask is designed to be swirled so that the sides are easily washed. Titrant is added from the buret until precisely the stoichiometric amount has been added. This point is usually signaled by the change in color of an <u>indicator</u>. The goal is to stop the addition of titrant just at the color change, or <u>end point</u>. Near the end of the titration, titrant is added a drop at a time to approach the end point cautiously. Partial drops of titrant are rinsed into the flask with a little distilled water from a wash bottle. With care, very accurate and precise titrations are possible.

The Titration

1. Obtain more reagent solution in your reagent beaker. Empty the buret and close the stopcock.
2. Fill the buret with reagent. Drain through the tip to remove air bubbles, then close the stopcock.
3. Refill the buret with reagent to above the 0.00 mL mark. Drain the buret into the spare beaker until the liquid level is just below the 0.00 mL mark (IT IS UNNECESSARY TO HAVE THE LIQUID LEVEL AT EXACTLY THE 0.00 mL MARK.) Touch the inside of the spare beaker to the buret tip to remove excess reagent from the tip. Read and record the volume level in the buret

to the nearest 0.01 mL. This is the initial buret reading.

4. Place an Erlenmeyer flask containing a sample to be titrated under the buret tip, and lower the buret so the tip is just above the lip of the flask. Open the stopcock and, swirling the flask, run reagent solution rapidly into the flask until you have delivered ≈ 80% of the volume estimated to reach the end point. Continue the titration, adding reagent "dropwise" as you swirl the flask. The endpoint color persists longer and longer as you near the end point. Stop at the first persistent color of indicator. If a drop of titrant remains on the buret tip, transfer it to the flask by touching the side of the flask to the buret tip. Rinse down the sides of the flask with distilled water. Read the liquid level in the buret. This is the final buret reading. If you are unsure that you have reached the end point, wash down the sides of the flask. Record the volume and add a fraction of a drop more by letting out a small amount of titrant and washing it from the buret tip into the flask. If the fraction of a drop clearly places you past the endpoint, use the previously recorded volume as the end point volume.

5. When finished, rinse the buret several times with distilled water, running some water out the tip each time. The buret should be full of distilled water and tightly clamped when you leave the lab.

Transfer Pipet

CAUTION: Never pipet by mouth. Use a pipet bulb or a syringe pipet pump (your instructor can show you how to make one of these). A transfer pipet has a bulge in the middle and a single calibration mark above the bulge. Its purpose is to deliver a single specific volume--an aliquot--of reagent to a receiver vessel. Common aliquot sizes are 1.000, 5.00, 10.00, 25.00, and 50.00 mL. See Table 1.

A. **Checking for Cleanliness**. To a clean beaker, add about 1 1/2 times more distilled water than the pipet will hold (e.g., about 40 mL of water for a 25 mL pipet). Fill the pipet to just above the calibration mark using a pipet bulb or pipet pump. Do not draw liquid up into the bulb or pump! Let the water drain into the sink by gravity after removing the bulb or pump. Examine the inner pipet wall from just above the calibration mark to the tip for water drops. No drops means the pipet is clean and you may rinse the pipet with reagent (see below). If drops appear, the pipet is dirty and should be exchanged for a clean one. Proceed only with a clean pipet.

B. **Rinsing with Reagent Solution**. Place about 40 mL of reagent solution into a clean dry beaker. Dry the outside of the pipet with a Kimwipe and hold the Kimwipe against the tip to absorb the water remaining in the tip. Use a pipet pump to draw solution into the pipet until liquid begins to enter the bulge of the pipet. Remove the pump and quickly place your right index finger over the top of the pipet. Tilt the pipet so solution flows up the inside of the pipet past the calibration mark, but not out the top. Rotate the pipet slowly to wet the inside surface with solution. Drain through the tip into a waste beaker. Rinse twice more.

C. **Pipetting a Reagent Solution**. Use the same beaker to obtain 1 1/2 times as much reagent solution as the total volume of the aliquots to be measured (e.g., if you want to pipet three 25 mL aliquots, get about 100 mL of solution). Use a pipet pump or pipet bulb to draw solution into the pipet to above the calibration mark. Remove the tip of the pipet from the beaker and wipe the lower portion of the pipet with a Kimwipe. Using the syringe pump or your index finger, lower the liquid level until the bottom of the meniscus is aligned with the mark. Touch the pipet tip to the inside of the beaker. Be careful not to transfer excess solution on the outside of the pipet. Place the tip of the pipet over the receiving vessel and let the pipet drain by gravity. Allow the liquid to drain from the tip until the flow stops. After 30 seconds, touch the tip of the pipet to the inner wall of the receiving vessel to remove the last drop of reagent. **NOTE: A small volume**

of reagent remains inside the pipet tip. This is normal--do not blow it out!

Measuring (Mohr) Pipet

A measuring pipet has no bulge in the middle and is graduated along its length like a buret. Its purpose is to deliver any convenient volume up to the maximum volume of the pipet. Common maximum volumes are 1, 5, and 10 mL. The uncertainty in a single reading of a measuring pipet is 1/10 of the smallest division of the pipet. For a 1-mL pipet graduated to 0.1 mL, a single volume reading has an uncertainty of 0.01 mL.

A. **Checking for Cleanliness**. This is done as for the transfer pipet. The measuring pipet should be clean from above the 0.00-mL mark to the tip.

B. **Rinsing with Reagent Solution**. This is the same as for the transfer pipet except that the measuring pipet should be filled about 1/3 full with rinse.

C. **Measuring out a Reagent Solution**. Use the same beaker used for the rinse solution to obtain a quantity of reagent solution about 1 1/2 times more than the total volume of the aliquots to be pipetted (e.g., 45 mL to pipet three 10-mL samples). Check the pipet for calibration marks. Some measuring pipets are graduated to the end of the tip while others have a bottom line where the meniscus must be stopped.

Using the pipet bulb or pipet pump, fill the pipet above the 0.00 mL mark. Remove the tip of the pipet from the beaker and wipe the lower portion of the pipet with a Kimwipe. Lower the liquid level until the bottom of the meniscus is aligned with the 0.00 mL mark. Touch the tip to the inside of the beaker. Place the tip of the pipet over the receiving vessel and lower the liquid level in the pipet with the pump or your index finger until the bottom of the meniscus is aligned with the desired mark. Touch the tip of the pipet to the inside of the receiving vessel, then remove the pipet. Note that some measuring pipets are calibrated to the tip. These are drained to deliver the total volume. Others are not calibrated to the tip and the liquid level must be stopped at the bottom calibration line. Carefully check a measuring pipet before using it.

Volumetric Flask

A volumetric flask is commonly used in the preparation of solutions of desired molarity. It has a single calibration mark on the neck with which the liquid meniscus must be matched. In matching the meniscus to the mark, you must avoid parallax error. Before preparing a solution in a volumetric flask, make sure that it is clean and that the stopper does not leak. The uncertainty in the volume contained by a volumetric flask is usually 0.1% of the flask volume. See Table 1.

A. **Checking for Cleanliness**. Rinse the volumetric flask several times with distilled water and examine the inner surface for drops. Drops indicate a dirty flask. Usually they appear in the neck. If drops appear, the flask should be exchanged for a clean one.

B. **Dilution of a Sample**. HANDLE THE FLASK BY THE NECK ABOVE THE MARK. Liquid samples are pipetted into the flask. However, solid samples are transferred quantitatively by dissolving in a small amount of solvent or with a solid funnel. Rinse the beaker directly into the flask insuring all of the solid is transferred to the flask. Use a wash bottle containing solvent to rinse down any sample on the inner surface of the neck of the flask above the mark. Half-fill the flask with solvent, and swirl to mix. DON'T splash solution above the mark. Add more solvent and mix well again. Bring the liquid level almost to the mark and allow time (10 sec) for drainage. THERE MUST BE NO DROPS OF LIQUID ABOVE THE MARK

WHEN THE LIQUID LEVEL IS BROUGHT UP TO THE MARK. Use an eye dropper to add solvent until the bottom of the meniscus is aligned with the mark. Insert the stopper firmly into the flask. Hold the stopper in with your thumb and shake the flask to mix.

Table 1
Volume Measurements in Common Laboratory Glassware

Glassware	Volume in mL	± Uncertainty	% of volume
volumetric	1.000	0.006	0.60
pipets	2.000	0.006	0.30
	5.00	0.01	0.2
	10.00	0.02	0.20
	25.00	0.03	0.12
	50.00	0.05	0.10
volumetric	10.00	0.02	0.20
flasks	25.00	0.03	0.12
	50.00	0.05	0.10
	100.00	0.03	0.03
	250.00	0.12	0.05
	500.00	0.20	0.04
buret	0.00-50.00	0.01	varies
Erlenmeyer	100	5	5
flasks	250	10	4
beaker	50	5	10
	100		5
graduated	10.0	0.1	1
cylinder	100.0	0.5	0.5

Finding the Equivalence Point on a pH Titration Curve

The equivalence point of an acid-base titration occurs when the moles of acid are equal to the moles of base. The progress of a titration may be followed by an indicator, a dye substance that changes color in the presence of an acid or base. An indicator is chosen such that the color change occurs at a pH near the equivalence point pH. For a strong acid/strong base titration, the equivalence point occurs at a pH of 7 so the indicator chosen will change color near pH 7. For a weak acid/strong base, the equivalence point pH will be greater than 7, somewhere between 8 and 10, so the indicator must change in the pH range 8 to 10. Finally, for a strong acid/weak base, the equivalence point will occur between a pH of 4 and 6 and the indicator must be selected accordingly.

Alternatively, the titration may be followed by monitoring the pH as the titrant is added as in Figure 1. Here a weak acid, acetic acid, is titrated by the strong base, NaOH. The initial pH is less than 7 and as the base is added, all of the OH^- reacts with the H^+ from the acid and more of the weak acid will dissociate. Therefore, the pH begins to increase slowly because there is less and less H^+ in solution. Finally, as the moles of base added approach the moles of acid initially present, small additions of base cause larger and larger

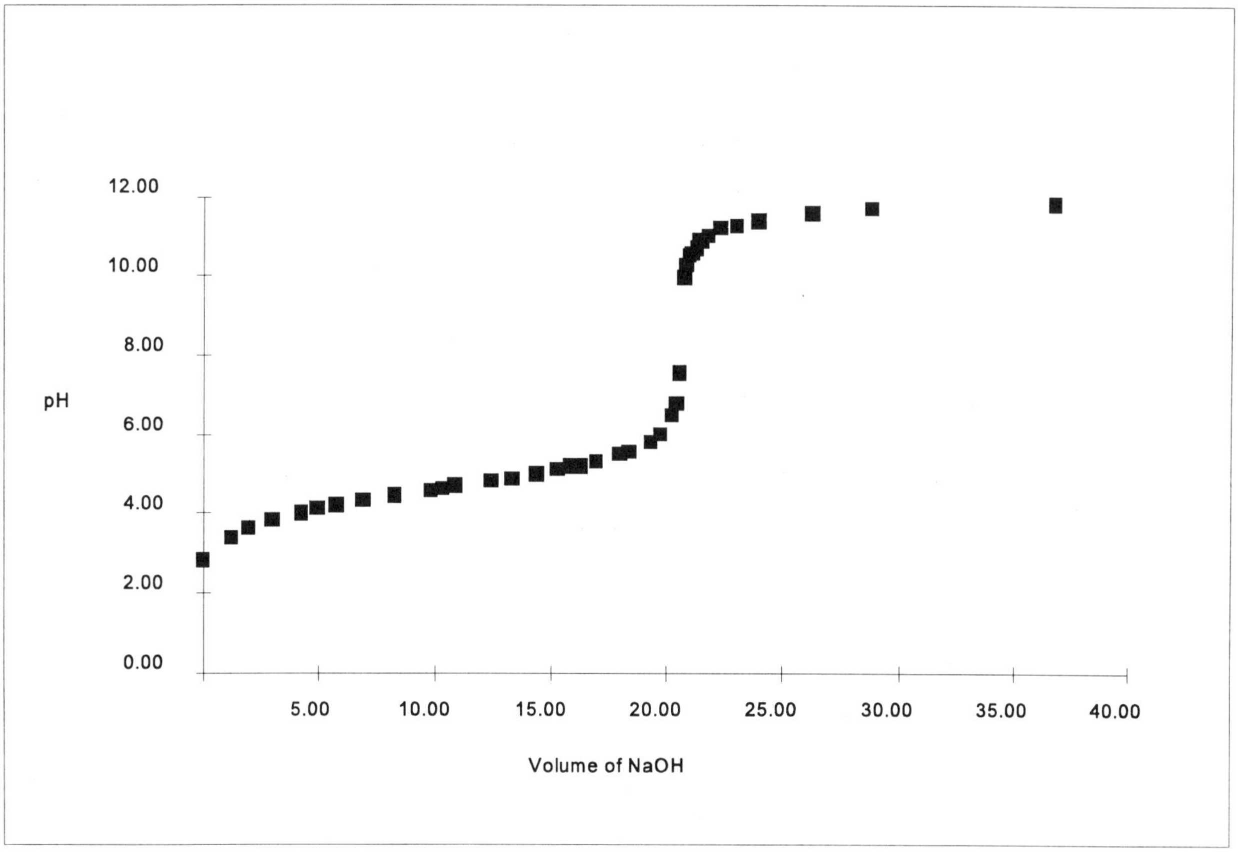

Figure 1. pH Titration Curve

changes in the pH. Near the equivalence point, the change in pH with the addition of base becomes sharp and the titration curve will appear almost vertical. Finally, as all of the acid is consumed, the further addition of base continues to increase the pH until it becomes approximately equal to the pH of the titrant and no longer changes with the addition of more base.

One method for locating the equivalence point is to use a graphical method. First, the titration curve is plotted as in Figure 1 above. However, the plot must have the same precision as the data, i.e., two decimal places on each axis. Usually, this precludes the use of a computer graph. Next, a line is drawn tangent to the flat portions of the titration curve, lines (A) and (B). See Figure 2. Note that these lines are generally **not** parallel to each other. Next, perpendiculars are drawn to the tangents. A perpendicular line is dropped from the upper tangent to the lower tangent and a perpendicular is drawn from the lower tangent to the upper, lines (C) in Figure 2. Again, these two lines are not necessarily parallel to each other. The midpoints on the perpendiculars are found and connected by a line (D). The point where the bisecting line (D) crosses the titration curve is the equivalence point. The equivalence point volume is found by dropping a perpendicular line from the intersection of the titration curve and line (D) to the x-axis (line E in Figure 2). Here the equivalence point volume is read from the graph. As seen in the example below, with a computer generated graph, the volume can only be estimated to the nearest milliliter and thus a hand drawn curve with greater resolution on the axes is necessary to use this method.

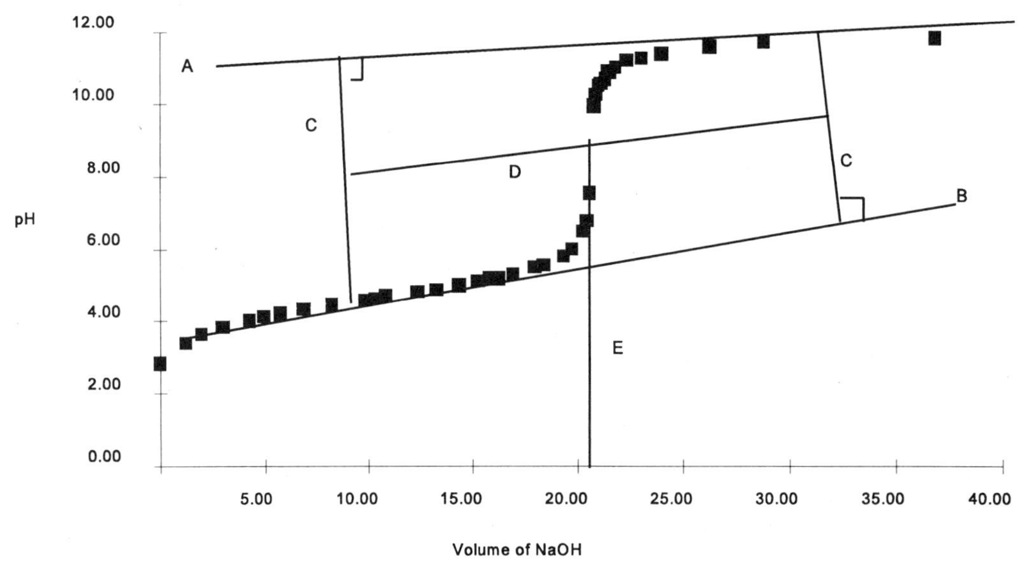

Figure 2. Graphical Method for Estimating the Equivalence Point

Another method for finding the inflection point on the titration curve is to use the first derivative. It is the inflection point of the sharply rising portion of the titration curve that defines the equivalence point. The inflection point on a curve occurs where the slope is at a maximum. Therefore, we can use the slope to find the equivalence point by finding the slope at each point on the titration curve near the equivalence point. The equivalence point occurs between 19 and 21 mL of titrant for the titration curve in Figure 2. To find the slope at each point of the titration curve take the difference in pH readings between two successive additions of titrant and divide by the difference in volume that produces the pH change. These calculations are easily carried out with a spreadsheet program such as Excel® or Lotus 123® and are illustrated below.

To use a spreadsheet, enter the volume of titrant added, corrected for the initial volume of titrant, into the first column, A. Note that it is not necessary to enter every data point, but only 15 to 20 points near the inflection point. In the next column, B, enter the corresponding pH values. Since we are interested in a plot of slope vs. volume of titrant, we must interpolate a value for the volume that corresponds to the new

	A	B	C	D
1	Volume of Titrant	pH	Plot Volume	Slope
2	mL		((vol2-vol1)/2)+vol1	(pH2-pH1)/(vol2-vol1)
3				
4	18.34	5.59	18.80	0.28
5	19.26	5.85	19.45	0.42
6	19.64	6.01	19.93	0.86
7	20.21	6.50	20.29	2.00
8	20.36	6.80	20.45	4.06
9	20.54	7.53	20.62	15.31
10	20.70	9.98	20.76	2.91
11	20.81	10.30	20.87	1.67
12	20.93	10.50	20.98	1.10
13	21.03	10.61	21.12	0.61
14	21.21	10.72	21.28	1.23
15	21.34	10.88	21.41	0.38
16	21.47	10.93	21.62	0.31
17	21.76	11.02	22.03	0.30
18	22.29	11.18	11.15	0.50

Figure 3. Data Entered to Spreadsheet

points of the slope. This volume is the midpoint between successive volumes of NaOH. First, calculate half the volume difference for successive additions of titrant and add the difference to the smaller volume. This value becomes the Plot Volume or the x -axis of the derivative graph (Equation 1).

$$\text{plot vol} = ((a5 - a4)/2) + a4 = ((19.26 - 18.34)/2) + 18.34 \qquad (1)$$

Here a5 and a4 refer to spreadsheet cells.

To determine the slope, calculate the change in pH and divide it by the change in volume between two successive points, as in equation 2 (b4 and b5 are also spreadsheet cells). Note that the volume and pH must correspond so that the slopes do not become negative.

$$\Delta pH/\Delta vol = (b5 - b4)/(a5 - a4) = (5.85 - 5.59)/(19.26 - 18.34) \qquad (2)$$

Resulting data may be plotted by hand or by the program, as in Figure 4. The maximum slope and corresponding volume represent the inflection point and may be used to approximate the equivalence point volume. In the example above, the maximum slope is 15.31 which corresponds to a titration volume of 20.62 mL of NaOH.

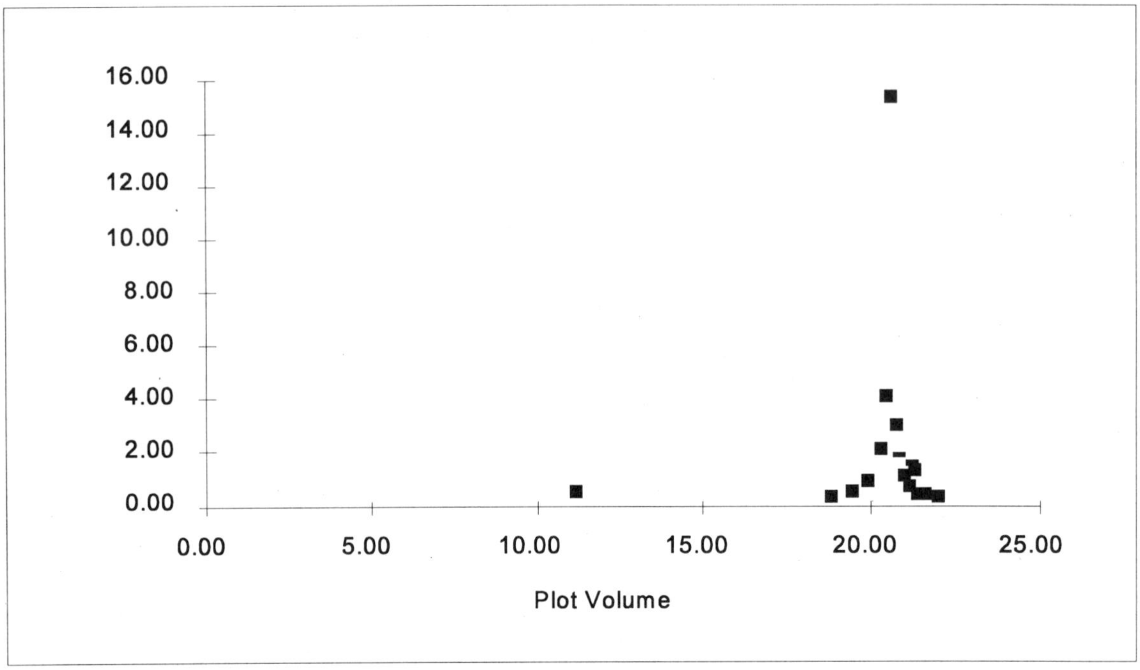

Figure 3. First Derivative of Titration Curve

Once the volume of base needed to reach the equivalence point is found, it is used to calculate the moles of the acid titrated. From the molarity and volume of the base, the number of moles of base is calculated:

$$(mol_{base}/L) \times 0.02062 \text{ L} = mol_{base} \qquad (5)$$

At the equivalence point the moles of base equal the moles of acid, so the moles of acid are known. From the volume of acid titrated, the molarity of the acid is calculated.

The Laboratory Notebook

The proper use of a laboratory notebook is an acquired skill requiring practice. However, it is a skill that will serve the student well in all endeavors, even if this is the only science course completed. The careful, complete recording of events and observations can be applied to any field. The keys to a good laboratory notebook are:

- good, concise descriptions
- clear layout
- good penmanship

Reasons for keeping the notebook are many. For example, useless repetition of experiments can be avoided by carefully recorded notes. In industrial situations, the notebook may be entered as evidence in patent disputes or of the thorough testing of products.

The notebook must be well organized and complete to serve as a useful tool. Students should spend time considering what information is important and what is the best way to organize that information. The **Results** page in the lab manual offers hints to help the student organize various experiments.

A laboratory notebook is a tool and not the goal of the laboratory work. The contents should be complete so that a reader with the same level of skills as the author could pick up the notebook and reproduce the experiments.

The Notebook

Bound notebooks are used to avoid loss of loose papers. Also, a permanently bound notebook avoids confusion in the ordering of experiments. Therefore, good practice precludes the removal of pages from the notebook or leaving blank pages. The exception here is the front of the notebook where a table of contents is entered as the work is completed.

All pages should be numbered consecutively. This allows the reader to find information easily. If the notebook used is not pre-numbered, take time to number the pages before beginning work. The numbers should be recorded in the upper outside corner of each page. A circle around the number helps eliminate confusion with any data recorded on that page.

A table of contents appears at the front of the notebook. Two to three pages should be left for this purpose. The table will include a concise list of the dates and experiments or projects performed, the pages where the information is recorded and the names of other group members. The table must be updated each time the notebook is used.

Proper Entry of Information

Experimental techniques, data, and observations are recorded as the work is performed. This helps eliminate losing vital information because the experimenter forgot a step when writing the procedure after the experiment was completed or lost scraps of paper with important data.

Clear, descriptive titles help the experimenter organize the important information. Titles also help the reader to follow the experiment and results. Proper grammar and spelling are important and the notebook should contain both. However, the goal is to record the experiment, not necessarily to write an award winning essay. As the student's skill in recording information improves, additional attention can be devoted to sentence structure and spelling.

Most commonly, the laboratory notebook is written in the first person. This allows the author to give credit to other individuals for their contributions to the work and to maintain a consistent format.

Occasionally, data will be obtained in the form of computer printouts or other output from instruments used. This information should be taped into the notebook as it is part of the experimental record. The instrument maker, model number, and settings used to record the data are also recorded in the notebook.

Writing Materials

Since the notebook is to be a permanent record of the work done, it is important that the writing materials be permanent also. For example, pencil is unacceptable because it can be erased and data can be changed or eliminated. Test ink before use to be sure it is waterproof. Spills are a common occurrence in lab and one spill can obliterate days' worth of work.

The use of correction fluids is not allowed because it destroys part of the permanent record. If an entry is in error, the entry is struck with a single line through it and the corrected entry made with a note about why the first entry was in error.

Organizing Information

The most important skill in keeping a good notebook is learning to organize the information so that it is easily retrieved and understood. Therefore, each experiment requires forethought and planning. Remember, for the notebook to be useful, someone must be able to read and understand what the experimenter did.

The format used for each experiment will vary depending upon the goals of the experiment. Students must learn to be flexible in the use and format of the notebook. For example, if the experiment is a series of titrations where the same procedure is used for each and the major emphasis of the experiment is to collect data, the notebook entry should reflect this with a single description of the procedure followed by a data table as in Figure 1. Here the student records the important information for each trial but does not record the steps followed because they are the same each time. Also, the calculations necessary to derive information from the data are included. The results are summarized in a second table, Figure 2. This table allows the data to be compared at a single glance. Deviations in the data are more obvious when compared in a table.

In another type of experiment, the student may be using an instrument or equipment for the first time. A complete description of the setup is necessary. In addition, a picture of the setup allows the experimenter to quickly reassemble the same apparatus with little effort. Subsequent experiments with the same equipment can refer to the previous drawings and descriptions.

Each experiment should begin on a new page. The title, date and partners are listed at the top of the page. Then the background information is given. This is the introduction to the experiment and includes why the experiment will be carried out and what the experimenter hopes to learn. Chemical equations pertinent to the experiment are given here. Sample calculations may also be included in the introduction.

The plan, or the experimental details for carrying out the study, follow the introduction. Note that if a procedure is outlined at this point, the actual steps in the experiment are still recorded as the experiment proceeds. To help eliminate confusion, plans are often outlined on the left-hand pages in the future tense. Actual experimental procedures are recorded on the right-hand pages in first-person passive voice, i.e., "I added 20 mL of distilled water to 5 g of the unknown in a 50 mL beaker." Along with the procedural steps, the observations and data are recorded. It is important that observations immediately follow each step. This helps the reader to understand what occurred in lab.

The final portion of each experiment is a study of the results. This includes a discussion of the results and any thoughts the experimenter has concerning the experiment. The discussion may include errors that occurred. A discussion of errors should include the source of the error and its likely influence on the results. It is understood that human error is unavoidable, and it should only be listed as an error if it can be identified. For example, in the titration example given previously, human error occurs whenever a buret is read. Therefore, it can be dealt with in an estimate of the uncertainty of reading a buret ($\pm$ 0.02 mL). However, if the endpoint of trial 1 was in error because the indicator color was wrong, this is a specific error with a specific effect on the data.

The conclusion is also part of the results. It is a brief, concise statement of what was accomplished or learned.

Good Lab Practice

When you set up an experiment, be sure to make room for the notebook and/or lab manual first. This will avoid having the notebook fall off the work bench or being the source of an accident. If you make room to write at the start of the experiment, recording information becomes less of a chore and will more likely become a good habit.

Be sure to note any safety precautions given either in this manual or by your instructor. Also, note any physical properties of the compounds you will be using. A simple table keeps this important information organized and easily located.

When recording data, use proper units and significant figures. Also, use good penmanship so your record is legible. Data that cannot be read cannot be used. The following important information should be included to make your records clear and concise:

- starting and stopping times for procedures such as heating, cooling and mixing
- colors, odors or changes such as formation of a gas or precipitate
- proper names of instruments or equipment used

Tables and Graphs

Both tables and graphs are important for organizing and presenting data. When preparing a table, consider the data to be recorded and leave space for calculated values and comments. Other points to remember:

- each table or graph requires a descriptive title
- number tables or graphs consecutively
- label the columns in tables or the axes on graphs with units
- note where any raw data are located in the notebook
- on a graph, use the same precision as the data, i.e., if the data are known to 0.01 g, then the graph should be to 0.01 g.

The workshop, **Graphing**, in this manual offers much useful information on graph construction.

Discussion or Evaluation of Data

A revised table of data, perhaps including calculated values, may be included here. However, the main purpose of this section is to record your thoughts about the experiment and what you perceive the results to be. You may include suggestions for improving techniques, equipment or amount of materials used. Include any errors that have occurred and possible consequences. For example, if you over-titrated because you are color blind, note this and the effect of over titrating, i.e., the standardized molarity is too small. Do not use the discussion to restate data; use it to express your understanding!

Conclusions

Restate the goal of the experiment and what was found. Does it support your hypothesis? What further experiments can be done either to support or refute what you have done?

Remember that keeping a laboratory notebook is an acquired skill nurtured by patience and practice and is an accomplishment worth striving to achieve!

Reference

Kanare, Howard M., **Writing the Laboratory Notebook**, American Chemical Society, Washington, D.C. 1985.

Figure 1. Sample from a Student Notebook

(13)

The titration of 0.1019 M KHP with 0.1 M NaOH.

The NaOH is being standardized.

Procedure:

Using a 25.00 mL volumetric pipet, add 25.00 mL of 0.1019 M KHP to each of 4 250 mL Erlenmeyer flasks. The pipet was rinsed 3 x with solution ~~in~~ before pipeting samples. Add 3 drops of phenolphthalein to the 1st flask, titrate to faint pink end point. Repeat 3 X with other samples. The data is summarized in Table 1.

Table 1. Titration data for the standardization of 0.1 M NaOH with 0.1019 M KHP

TRIAL	Vol KHP (mL)	Final vol NaOH (mL)	Initial Vol NaOH (mL)	Total vol NaOH (mL)	comments
1	25.00	35.73	10.46	25.27	v. light pink
2	25.00	30.62	5.32	25.30	good end point
3	25.00	37.91	12.66	25.25	light pink - good
4	25.00	36.30	11.02	25.28	yeah! good endpoint

Next, calculate the molarity of the NaOH from this data.

Figure 2. Sample from a Student Notebook

(14)

To calculate the molarity of NaOH to 4 sig. figs., the number of moles used is calculated. Since moles acid = moles base at the end point,

$$\text{moles KHP} = 0.02500\,L \times 0.1019\,\frac{mol}{L} = \text{moles NaOH}$$

$$\text{moles NaOH} = 2.5475 \times 10^{-3}$$

$$\text{Molarity NaOH} = \frac{2.5475 \times 10^{-3}\,mol}{\text{Volume used}}$$

For trial 1, Volume used = 25.27 mL = 0.02527 L

$$\text{Molarity} = \frac{2.5475 \times 10^{-3}\,mol}{0.02527\,L} = 0.1008\,M$$

Results are summarized in table 2.

Table 2. Summary of titration data for standardizing NaOH with KHP.

TRIAL	VOL NaOH used (mL)	moles NaOH $\times 10^3$	molarity NaOH mol/L
1	25.27	2.5475	0.1008
2	25.30	2.5475	0.1007
3	25.25	2.5475	0.1009
4	25.28	2.5475	0.1008

Average molarity of NaOH 0.1008 mol/L

SD ±0.0001 mol/L

Non-Instrumental Experiments

Fundamentals†

1 lab period; work in pairs. Complete the Preparation page before laboratory.

Goals

- To learn the operation of the electronic analytical balance
- To learn how to use the buret
- To learn about experimental uncertainty
- To practice hypothesis building and testing

Background

The purposes of this experiment are to introduce you to the use of the analytical balance, one of the most useful tools of the chemist, and to answer the following question:

What happens to the mass of a penny as it ages; does the mass increase, decrease, or remain about the same?

Equipment and Materials

analytical balance
50-mL buret
50-mL graduated cylinder
10 pennies

Safety

The materials used in this experiment are not harmful. Nonetheless, you should wear your safety goggles at all times in the laboratory.

Experimental, Part 1 (Part 2 of the Procedure will be given out during lab).

Record all data either in your laboratory notebook or on the **Results** sheet at the end of the experiment, whichever is applicable. Use of the analytical balance and operation of the buret will be demonstrated in class. You will find burets set up in the lab, filled with distilled water and capped.

Each team should obtain 10 pennies from the instructor. Determine the mass of each penny, and record mass and year. Repeat the procedure so that you have two determinations of the mass of each penny. Calculate the average mass for each penny, and the precision (1/2 the difference of the two masses). Your instructor will indicate how you are to handle your penny data.

Obtain a 50-mL graduated cylinder from the instructor. Clean and dry it. Obtain a buret, rinse it with distilled water as instructed in **Use of Volumetric Glassware** in this manual, and fill it with distilled

---

†Adapted with permission from the "Pennies" experiment developed at the College of the Holy Cross, copyright RonJon Publishing, Inc., 1993.

water to near the 0.0-mL mark. Record the initial volume reading to the nearest 0.01 mL. From the buret, carefully fill the graduated cylinder with water to the 45-mL mark, and record the final buret reading. Empty and dry the graduated cylinder, refill the buret, and repeat twice more. When finished, work on the **Questions** while waiting for discussion.

Disposal Methods

No disposal required.

Questions

1. Do your mass data support or refute the hypothesis developed at the beginning of the lab period?
2. Do your mass data exhibit any anomalies (aberrations)?
3. Are the anomalies real or are they due to error in your use of the balance?
4. What explanation can you offer for the anomalies?

References

1. "A Laboratory-Centered Approach to Teaching General Chemistry" by Robert W. Ricci and Mauri A. Ditzler. *Journal of Chemical Education*, **1991**, 68, 228-231.
2. "An Experiment in Thinking Scientifically" by Dennis J. Sardella. *Journal of Chemical Education*, **1992**, 69, 933.

Preparation
Fundamentals

Name ____________________
Section____________________
Date ____________________

Read

1. **Determining Mass**, in this manual.
2. **Use of Volumetric Glassware**, in this manual.
3. **Uncertainty, Precision, and Error in Experimental Results**, in this manual.
4. The appropriate sections of your textbook.

Problems

1. Two people find the mass of the same object. They obtain masses of 3.1101 and 3.1120 g, respectively. What is the uncertainty in each value? What is the precision in the mass of the object?

2. An object has a total mass of 16.0402 g and a volume of 2.00 mL, measured by water displacement. What is the density of the object? What is the precision in the density?

3. A five-cent piece has a mass of 4.0245 g. What is its volume? (The density of nickel is 9.91 g/mL.)

Pre -1980

no pennies

3.0 - 23.20 = 20.20

w/ pennies

3.61 - 19.05 = 15.44

4.76 val

$\frac{30.71}{4.76}$ (mass) = 6.452 gr.ml

Post 1980 -

no pennies

3.90 - 24.10 = 20.20

w/ pennies

4.0 - 20.00 = 16.00

4.2 val

$\frac{25.425}{4.20}$ = 6.053 g.ml

Conclusion:

low because of air bubbles

Density = 8.92

Pre 1980	Post 1980
9.614	7.453
8.138	6.531
7.161	5.00
8.38	7.21
7.702	7.34
6.452	6.35
8.107	6.05

Read Pg. 209

~~John~~ Dub. 454-1924

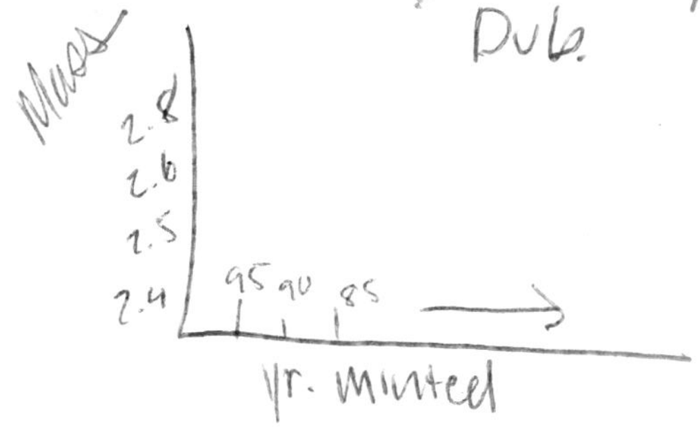

Results
Fundamentals

Name Nick

Section ______

Coworker John

Date ______

Data

Wt. <1980 = 30.710 Pre
>1980 = 25.425 After

Procedure, Part 1

Penny mass

Penny #	Year	Mass 1	Mass 2	Avg. Mass ± Precision
1	1983	2.510	2.510	2.510
2	1972	3.092	3.090	3.091
3	1977	3.080	3.080	3.080
4	1965	3.040	2.986	3.013
5	1971	3.051	3.050	3.050
6	1993	2.480	2.420	2.450
7	1973	3.043	2.980	3.011
8	1964	3.067	3.066	3.066
9	1979	3.080	3.080	3.080
10	1994	2.50	2.440	2.47

Volume of distilled water needed to fill graduated cylinder to 45-mL mark

Trial 1: final buret reading ______ mL
initial buret reading ______ mL
volume dispensed ______ mL

Trial 2: final buret reading ______ mL
initial buret reading ______ mL Average volume ____±_____mL
volume dispensed ______ mL

Trial 3: final buret reading ______ mL
initial buret reading ______ mL
volume dispensed ______ mL

Answers to Questions

1st reading Buret
~~1.95~~ ~~38.2~~
~~35~~ ~~38~~

Buret
3.00 23.20 of H_2O no pennies 20.20
3.61 19.00 old w/ pennies 19.00
4.20
new
3.90 24.10
4.00 3. 20.00 new

Nuts and Bolts

1 lab period; work in pairs. Complete the Preparation page before laboratory.

Goals

- To find the mass of a bolt that has a nut attached to it
- To determine the ratio of nuts to bolts in a sealed vial

Background

Frequently during an experiment, some quantity of interest cannot be measured directly. Therefore, an indirect method is developed to measure the quantity. For example, the mass of an electron can only be found indirectly from the mass to charge ratio as the mass of the electron is too small to measure directly. In this experiment, you will find the mass of a bolt by indirect methods. Each group has a bolt with a nut attached. The nut may not be removed from the bolt during the experiment. After using at least 2 methods to estimate the mass of the bolt, you will decide which method gives the better estimate of the mass by considering the errors in each method.

When making measurements, you need to consider what errors may occur, and the intrinsic error due to the instruments or equipment used. For example, if a double beam balance is used in this experiment, it will have an absolute error in each reading of $\pm$ 0.02 g. (If using a top loading or analytical balance, check with the instructor as to the uncertainty in the mass.) Therefore, if the mass of a nut is 3.00 grams, the uncertainty in this mass is $\pm$ 0.02 g and the true mass is most likely between 2.98 g and 3.02 g. Thus, in this experiment, the absolute uncertainty is determined by the measuring device, the balance.

The estimate of the true mass of any one nut can be improved by making multiple measurements and averaging the masses. However, generally one determination of the mass is adequate. Alternatively, the mass of a large number of nuts may be determined and averaged. If greater accuracy is required than $\pm$ 0.02 g, a balance with more precision, i.e. decimal places, may be used.

In this experiment, you are not able to measure the mass of the bolt directly because it is attached to a nut. Therefore, you must estimate the mass of the bolt using two different methods. Each method will have a precision associated with it and you should estimate the precision of each method. One way to estimate the reproducibility or precision of a measurement is by the standard deviation. Another method is to use one-half of the range. The standard deviation or one-half the range can then be used to help define the precision of a measurement along with the absolute uncertainty in the equipment.

The second part of the experiment uses the data from the first part, the mass of a nut and the mass of a bolt. Your group will receive a sealed vial. Given the mass of the empty vial, you will determine the ratio of nuts to bolts in the vial within the uncertainty of the data available.

How do you determine a mass when you cannot directly weigh the object of interest?
Does the method used affect the uncertainty of the measured value?

Equipment and Materials

lab glassware: beakers, graduated cylinders, Erlenmeyer flasks
balance - double beam, top loading or analytical
bolt with nut attached

bolts
nuts
sealed vial with unknown ratio of bolts and nuts

Safety

Eye protection must always be worn in the laboratory. The materials used in this experiment are not harmful, however, good lab practice requires the use of eye protection.

Experimental

Proper use of the balance will be demonstrated in lab. Each team will be given a bolt (B) with a nut (N) attached, and a supply of nuts and bolts similar to the nut-bolt combination. Devise at least 2 methods for estimating the mass of the bolt without removing the nut. Before you begin, clearly explain in one or two sentences what method you are using and what data you will need to obtain. Then, prepare a table to record trial numbers and the data obtained. If the mass is to be used in a calculation, such as subtraction, leave room in the table for the calculated values as well.

You and your partner will be given a sealed vial containing an unknown combination of nuts and bolts and the mass of the empty vial. From the information obtained in part 1 of this experiment, determine the ratio of nuts to bolts in the vial.

Disposal Methods

Return all bolts, nuts and vials to the instructor.

Questions

1. Report the two values for the mass of the bolt that you have found from your two methods. For any measured or calculated values report a standard deviation or one-half range, if appropriate.
2. Discuss the precision of your results in terms of the standard deviations. Do you think one of your methods gives a better estimate of the "true" mass of the bolt than the other? Why or why not? How could you decide which method is more accurate?
3. Report the ratio of N to B in your unknown combination of nuts and bolts. Explain the logic followed in deciding this ratio. How does knowledge of the uncertainty in the mass of the nut and/or bolt affect your conclusions?
4. Can you think of other ways to carry out this experiment now that you have completed it?

Preparation
Nuts and Bolts

Name ____________
Section ____________
Date ____________

Read

1. This experiment.
2. **Uncertainty, Precision, and Error in Experimental Data** in this manual and your text.
3. Dalton's Law of Multiple Proportions in your text.

Questions

Define the following terms:

accuracy
precision
systematic error
random error
Law of Multiple Proportions

Practice Problems

Student 1 measured the mass of 5 nuts and recorded the following data in her notebook:

Nut number	Mass in grams		
1	3.62		
2	3.67	average mass of a nut	______
3	3.70	standard deviation	______
4	3.96	one-half the range	______
5	3.62		

Using the equations given in the section on **Uncertainty, Precision, and Error in Experimental Data**, calculate the average mass of a nut based on this data. Calculate the standard deviation and one-half the range based on the data. Remember that n is equal to the number of data points used to calculate the average.

Student 2 obtained an average mass of 3.80 g with a standard deviation of 0.10 g. Do you think this average is significantly different from the average obtained by student 1? Why or why not?

Suppose student 3 obtained an average mass of 3.40 g with a standard deviation of 0.12 g. Should the students question this result? Why or why not? Which average is more precise? Why?

Results
Nuts and Bolts

Name ______________
Coworker ______________
Section ______________
Date ______________

Data

Part 1:
Description of first method:

Table 1. Data to determine the mass of the bolt - method 1

Description of second method:

Table 2 Data to determine the mass of the bolt - method 2

Part 2:

Table 3. Data to determine the ratio of nuts to bolts in the unknown

Vial number:	______
Mass of the vial:	______
Mass of the nuts & bolts	______
Number of bolts	______
Number of nuts	______

Explain the reasoning used to determine the number of nuts and bolts:

Answers to Questions

Volumetric Measurements and Solutions

1 lab period; work individually to calibrate a pipet; work in groups to prepare solutions. Complete the Preparation page before laboratory.

Goals

- To determine the accuracy and precision of a 25 mL volumetric pipet
- To find the density of an unknown solution using two different types of glassware
- Prepare a solution of assigned concentration with specified precision

Background

Experimental procedures frequently require measurements of specific volumes of liquids or mixing of particular concentrations of solutions. The type of balance and glassware most appropriate for measuring and transferring these solutions will vary, depending on the number of significant figures required. For example, to prepare 1 L of 0.01 M NaCl requires the mass of NaCl to be known to one significant figure (0.6 g) and to be diluted to approximately 1 liter with water (in a beaker). To prepare 1.000 L of 0.01000 M NaCl, on the other hand, requires the mass of NaCl be known to the nearest 0.0001 g (four significant figures), transferred quantitatively to a volumetric flask, and diluted to the mark to be sure that the total volume of the solution is 1.000 L (four significant figures). Review the data in **Use of Volumetric Glassware,** Table 1. Note that volumetric pipets, volumetric flasks and burets are used when precision (three to four significant figures) is required.

The accuracy of measurement with various types of laboratory glassware depends on two factors: 1) the skill of the person doing the measurement, and 2) the accuracy in manufacture of the glassware used. Accuracy of a variety of pieces of laboratory glassware is given in **Use of Volumetric Glassware**, assuming measurements are made and read correctly by the experimenter.

In this experiment, you will be testing both the accuracy and precision of the pipet in transferring a specified volume of solution. The manufacturer lists the accuracy as 25.00 $\pm$ 0.03 mL. Thus, if the pipet is used correctly each time, the volume transferred will be 24.97 to 25.03 mL or within 0.12 % of the desired value. Since using a pipet requires practice, the skill of the experimenter will have an effect on the accuracy. Moreover, the skill of the experimenter will affect the precision. Recall that accuracy refers to how closely data agree with the "true" value while the precision refers to how reproducible an experimental value is. You can estimate the accuracy of each measurement by comparing the volume transferred to 25.00 mL. The standard deviation is a measure of how reproducible your results are. Note that it is possible for results to be precise, very reproducible, but not accurate.

How accurately can you transfer 25.00 mL of solution?
How precisely can you transfer 25.00 mL of solution?

Equipment and Materials

25.00 mL volumetric pipets
weighing bottles
analytical balance
unknown liquids of varying densities

solid $CuSO_4$
solid NH_4NO_3
ethanol, C_2H_5OH
14 M NaOH

Safety

Eye protection must always be worn in the laboratory. The 14 M NaOH is very corrosive. Use gloves when handling the solution. Immediately wipe up any spilled chemicals with water. Note that base will attack glass. Thoroughly clean any glassware with plenty of water to prevent etching by caustics.

Experimental

Before beginning the experiment, briefly state in your notebook or on the **Results** page what approach you will use. For procedures that are repeated each time, e.g., drawing solution into the pipet and transferring it to a container, write the procedure once then compile the data in a table. Leave room for values to be calculated from the raw data, such as the volume of the pipet.

Part 1 (work individually)

Determine the "volume" of a 25.00 mL pipet. Use the density of H_2O at room temperature to devise a scheme to test the accuracy of the volume of a 25.00 mL volumetric pipet. Report the "volume" of the pipet you have used. Calculate the standard deviation of this value.

Find the density of an unknown liquid using two different types of glassware. Use the 25.00 mL pipet as one measuring device. Then determine the density of a second 25 mL sample with a second type of glassware. Pay attention to significant figures in reporting values. Report the standard deviation for each measurement.

Part 2 (work with a partner)

You will prepare an assigned solution from the list below as a demonstration for the class. Be specific and detailed in describing the amounts of solute and solvent used, measuring devices, glassware, mixing techniques, and any other observations. Calculate the concentration of your solution to the correct number of significant figures based on the actual amounts of reagents used.

a. 25 mL of a 0.1 M solution of copper sulfate, $CuSO_4$.

b. 25.00 mL of a 0.5 M solution of $CuSO_4$ to four significant figures. NOTE: the solution need not be EXACTLY 0.5 M, but must be known to 4 SIGNIFICANT FIGURES.

c. 100.00 mL of a 0.3333 M solution of ammonium nitrate, NH_4NO_3. See NOTE in b above.

d. 250 mL of a 0.7 M solution of ethanol, C_2H_5OH, in water without weighing anything. The density of ethanol at 25 °C is 0.80 g/mL.

e. an ethanol solution by pipeting 50.00 mL of H_2O into a volumetric flask followed by pipeting 50.00 mL of ethanol. What is the final volume? Can you calculate the molarity of the solution?

f. 1 L of 0.1 M NaOH from 14 M NaOH.

Disposal Methods

Copper solutions - dispose of in beakers or bottles labelled "Cu waste".

Solutions - wash down the drain with water.
Leave bench tops and balance areas clean. Return the rinsed volumetric flasks and pipets.

Questions

1. How is the precision of your average pipet volume affected by your skill in using the pipet?
2. How could you reduce errors in your results caused by the experimenter?
3. How important are multiple measurements?
4. What other sources of error might cause your calculated value to deviate from 25.00 $\pm$ 0.03 mL?
5. Report a density value $\pm$ standard deviation for the unknown liquid for each method. Do they differ? Why or why not? Which is more precise? More accurate? What factors affect the final values of density?

Preparation
Volumetric Measurements and Solutions

Name _______________
Section _______________
Date _______________

Read

1. This experiment.
2. Sections in text on preparing solutions.
3. **Use of Volumetric Glassware** and **Determining Mass** in this manual.

Questions

Define the following terms:

significant figures
solute
density
solvent
molarity

Explain the difference between "to deliver" and "to contain."

Look up the density of H_2O in the CRC Handbook of Chemistry and Physics or Lange's Handbook.

Practice Problems

1. Perform all of the operations and report the answers with the correct units and number of significant figures and units.

a. $\dfrac{0.1045 \text{ mol/L} \times 0.05000 \text{ L}}{0.02500 \text{ L}}$ = _________

b. A student used the "weigh by difference" method to obtain a sample of NaCl. From the data recorded calculate the amount of NaCl used:

mass of NaCl + weighing bottle	25.6154 g
mass of NaCl + weighing bottle	20.4455 g
mass of NaCl transferred	_________

c. The NaCl above was dissolved in 100.00 mL of H_2O. What is the molarity of the resulting solution?

d. 0.050 atm + 0.101 atm = _________

2. A 50.0 mL sample of pure acetic acid (the acid found in vinegar) had a mass of 52.46 g. What is the density of acetic acid?

3. 2-propanol (also called isopropanol and found in rubbing alcohol) has a density of 0.7855 g/mL. A 6.53 g sample of 2-propanol occupies what volume?

4. A 2.0436 g sample of potassium hydrogen phthalate ($KHC_8H_4O_4$, a primary standard) was quantitatively transferred to a 100.00 mL volumetric flask and water was added to the mark. What is the molarity of this solution?

5. How many moles are needed to make 52 mL of 0.0056 M NaOH?

6. Describe how you prepare 5.00×10^2 mL of 0.02 M $CuSO_4$. How many grams of $CuSO_4$ are needed? How many significant figures are used?

7. Describe EXACTLY how you would prepare your assigned solution from the list of solutions in **Part 2** of the **Experimental** section. Include glassware and equipment needed.

Results
Volumetric Measurements and Solutions

Name ____________
Coworker(s) ____________
____
____
Section ____________
Date ____________

Data

Part 1:

Experimental approach:

Table 1. Volume of 25.00 mL pipet

Trial Number	Mass of Bottle	Mass of Bottle + H_2O	Mass of H_2O	Volume of H_2O

Average volume ________
Standard deviation ________

Determination of the Density of an Unknown

Unknown ______ Density__________

Experimental approach:

Table 2. Density of unknown with 25.00 mL pipet

Trial Number	Mass of Bottle	Mass of Bottle + unknown	Mass of unknown	Volume of unknown

Average density__________
Standard deviation________

Determination of the density of an unknown with a second type of glassware

Table 3. Density of unknown with________________

Trial Number	Mass of Bottle	Mass of Bottle + H_2O	Mass of H_2O	Volume of H_2O

Average density__________
Standard deviation________

Name ________________

Part 2:

Preparation of solution(s)

Assigned solution(s) _______

Procedure and observations

Answers to Questions

Acid-Base Titrations

1 lab period; work in pairs on Part 1; individually on Part 2. Complete the Preparation page before laboratory.

Goals

- To learn to make accurate and precise volumetric measurements
- To titrate an unknown acid and plot titration curves
- To find the molarity of an unknown acid

Background

Titration is an analytical technique used to find the exact concentration of a solution to at least four significant figures. The equivalence point of a titration occurs when the moles of reactant in the titrant are equal to the moles of reactant in the sample. To visualize when the end point occurs an indicator is added. Indicators are dyes, weak acids, that change color in the presence of a very small excess of acid or base. For this acid-base titration, the standardized base, NaOH, will be added to an unknown acid, and followed by an appropriate indicator. Thus, the endpoint of the titration is signaled by the color change observed.

Indicator titrations are carried out in Erlenmeyer flasks that can be swirled for good mixing of reagents. Also, the sides can be washed down easily so no materials are lost on the sides of the flask. Your instructor will demonstrate proper titration technique in the lab.

For a pH titration, a small beaker (100 or 150 mL) is used. The pH electrode and a mechanical stirrer are immersed in the solution to be titrated. The pH electrode is not removed as the titrant is added. Titrant is added in small increments of 0.5 to 1 mL, and the pH and volume are recorded after each addition. As the equivalence point is approached, the pH begins to change more rapidly with each addition of titrant and the volume of titrant added each time should be decreased. Once the equivalence point is past, the pH change with addition of titrant slows again and larger increments of titrant can be added until the pH is approximately constant. See **Finding the Equivalence Point on a pH Titration Curve** in this manual for data treatment.

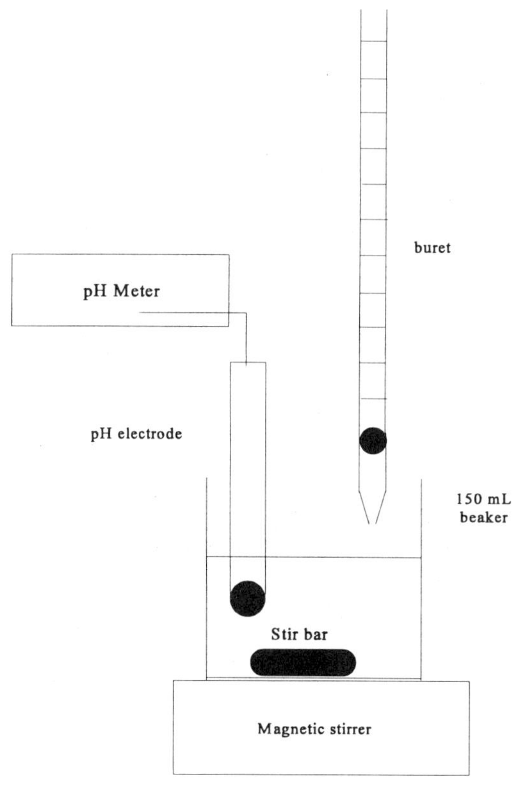

Figure 1. Experimental arrangement of a pH titration.

You should plan a titration such that the volume of titrant added is between 20 and 30 mL. By planning to use less than a full buret (50 mL) you can avoid running out of titrant and having to repeat the measurement. To have four significant figures, the volume of titrant must be greater than 10.00 mL.

Does an indicator titration yield the same results as a pH titration?
If not, how are they different?

Equipment and Materials

50 mL buret
10, 15, 20, 25 mL volumetric pipets
Erlenmeyer flasks
pH meter
magnetic stirrer and bar
0.1 M NaOH (standardized)
standard buffers 4, 7, and 10
unknown acid A and indicator A
unknown acid B and indicator B

Safety

Eye protection must always be worn in the laboratory. Dilute acids and bases will cause irritation if left on the skin . Immediately wash spills with soap and water. Bases attack glass, so rinse the buret and fill with distilled water when you are finished.

Experimental

Record all data either in your laboratory notebook or on the **Results** page at the end of the experiment, as directed. Practice reading the burets set up at the front of the lab. Do not waste time standing in line to read the burets; come back when they are free. Record the liquid level in each and the difference between volumes on the board.

Part 1: pH titration of an acid (work in pairs)

Calibrate the pH meter as directed. Titrate your unknown acid using a pH meter. Record the volume of NaOH added and the pH at least every 0.5 -1 mL initially and more frequently when the pH changes more rapidly with volume. Titrate to an approximately constant pH greater than 10. Plot pH vs. volume of NaOH added. The equivalence point volume is found at the inflection point of the curve. See **Finding the Equivalence Point of a pH Titration** in this manual. Calculate the molarity of the unknown acid.

Part 2: Titration with an Indicator (work individually)

Prepare your buret as demonstrated by the instructor and described in this manual. See **Use of Volumetric Glassware**. Your initial reading should not be 0.00. Determine the color change for your indicator. Perform the titration with standardized NaOH and your unknown acid. How much acid should you titrate? Consider how you will measure the unknown acid recalling you need four significant figures. Record all data in an appropriate table.

Calculate the molarity of your unknown acid from your data. Estimate your precision using a standard deviation. Before leaving lab, exchange data with a group using the other unknown acid. Plot the pH vs. volume of NaOH for this second set of data.

Disposal Methods

The solutions of acids and bases may be flushed down the drain with excess water unless otherwise directed by your instructor.

Questions

1. Calculate the molarity of your unknown acid to the correct number of significant figures for each titration method used. Include an estimate of precision. Are the calculated molarities the same for each method?
2. Analyze the titration curves by finding the equivalence points of the titrations. Report the volume of NaOH needed to reach the equivalence points. Does this volume correspond to the volume used for the indicator titrations? Explain.
3. Discuss any differences observed between the titration of unknown acid A and unknown acid B.
4. Summarize your results in a table.
5. Comment on precision and accuracy of your experimental work. How could you improve either or both?

Preparation
Acid-Base Titrations

Name ____________________
Section ____________________
Date ____________________

Read

1. This experiment.
2. **Use of Volumetric Glassware** and **Finding the Equivalence Point of a pH Titration Curve** in this manual.
3. Assigned sections of the text.

Questions

Write balanced chemical equations for the following reactions. Both are neutralization reactions.

____NaOH + ____ HCl →

____NaOH + ____ H_2CO_3 →

Define the following terms:

neutralization
acid
base
pH
standardizing a solution
net ionic equation
balanced chemical equation
spectator ions

Practice Problems

1. For the following reaction, calculate the volume of 0.1503 M HCl that reacts with 20.00 mL of 0.1985 M sodium carbonate according to the reaction:

$$2HCl(aq) + Na_2CO_3(aq) \rightarrow 2NaCl(aq) + H_2O(l) + CO_2(g)$$

2. What volume of 0.2515 M sulfuric acid reacts with 50.00 mL of 0.1150 M sodium bicarbonate?

$$H_2SO_4(aq) + 2NaHCO_3(aq) \rightarrow Na_2SO_4(aq) + 2\ H_2O(l) + 2CO_2(g)$$

Results
Acid-Base Titrations

Name ____________
Coworker(s) ____________
Section ____________
Date ____________

Data

Trial buret readings:

Initial volume____________

Final volume____________

Total volume____________

Part 1. Titration with a pH meter

Table 1. pH Titration Data for the Titration of Unknown with NaOH

Trial 1			Trial 2		
Buret Reading	Volume NaOH used	pH	Buret Reading	Volume NaOH used	pH

Part 2. Indicator titration of unknown acid:

Unknown acid used______________

Volume of acid used______________

Molarity of acid used______________

Table 2. Titration of unknown acid with NaOH

Trial number
Final buret readings
Initial buret reading
Total volume NaOH
Calculated Molarity of unknown

Answers to Questions

Atomic Weight

1 lab period; work in pairs. Complete the Preparation page before laboratory.

Goals

- To discover the connection between electric current, electrons, and oxidation states of elements
- To apply the ideal gas law
- To determine the atomic weight of an element

Background

Electrolysis is used for the production of several important materials on a huge scale. Aluminum, sodium, chlorine, and sodium hydroxide are produced in this way, and copper is purified electrolytically.

In this experiment you will construct an electrolysis cell as shown in Figure 1. The cell consists of a beaker containing an aqueous solution of 0.1 M acetic acid, $HC_2H_3O_2$, and 0.5 M sodium sulfate, Na_2SO_4, which is present to conduct electricity through the solution. Immersed in the solution are a strip of zinc metal and one end of a copper wire. The copper wire and zinc electrode are connected through a 1.5 V flashlight battery with the + terminal connected to the zinc electrode. The battery pushes electrons out of its negative terminal toward the submerged end of the copper wire, while the positive terminal of the battery pulls electrons from the zinc electrode. Zinc atoms become oxidized to Zn^{2+} at the zinc electrode, and the zinc ions dissolve into solution. The zinc electrode is called the **anode**, the electrode where oxidation occurs. Electrons from the battery move to the Cu wire and react with H^+ ions in the solution, reducing them to $H_2(g)$. Hydrogen gas rises into the closed end of the graduated cylinder, where it is trapped. The copper wire is called the **cathode**, the electrode where reduction occurs. The reactions occurring in the cell are thus

Figure 1

Electrolysis Setup

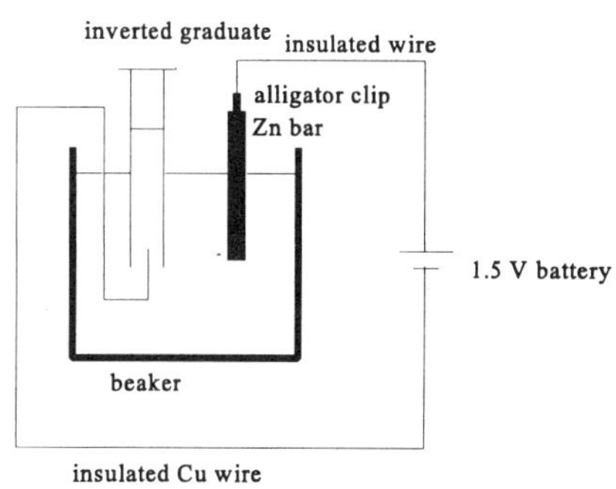

$$Zn(s) \rightarrow Zn^{2+}(aq) + 2e^- \qquad \text{at the anode}$$

$$2H^+(aq) + 2e^- \rightarrow H_2(g) \qquad \text{at the cathode}$$

$$Zn(s) + 2H^+(aq) \rightarrow Zn^{2+}(aq) + H_2(g) \qquad \text{overall}$$

The volume of hydrogen gas collected will be studied as a function of the mass of Zn lost from the anode during electrolysis.

While the electrolysis is proceeding, you will carry out a second experiment, in which you will react Zn metal with an excess of strong acid, and collect the hydrogen gas produced. In this case, you will measure the volume of hydrogen gas as a function of the mass of Zn used. The purpose of your measurements is to answer the following questions:

How can data of this type be used to determine the atomic weight of Zn?
How do the results from the two methods compare?

Equipment and Materials

150-mL beaker
6-inch test tube
thermometer
test tube rack
insulated electrical wires
alligator clips
electrical tape
10-mL graduated cylinder
1.5 Volt flashlight battery (size D)
stirring rod
25-mL graduated cylinder (for collecting gas)
2-foot length of tygon tubing
sandpaper or carborundum paper
50-mL glass syringe
forceps or tongs
#2 1-hole rubber stopper with 2-inch length of glass tubing
clamps
analytical balance
glass wool
Zn electrode
6 M HCl
0.1 M $HC_2H_3O_2$ in 0.5 M Na_2SO_4

Safety

Safety glasses must be worn at all times in the laboratory. You will work with solutions of hydrochloric acid, acetic acid and sodium sulfate. Avoid ingestion of solution. Hydrogen gas will be generated during the experiment. NO FLAMES PLEASE!

Experimental

Record all data in your notebook or on the **Results** sheet, whichever is applicable.

Electrolysis.† You should set up and run two electrolyses at the same time. For each electrolysis set-up, clean the 25-mL graduated cylinder and beaker, and rinse with acetic acid/sodium sulfate solution (electrolyte). Fill the beaker with electrolyte, then fill the graduated cylinder completely from the beaker.

---

†Adapted with permission from Hunt and Block, Laboratory Experiments for General Chemistry, 1990, copyright Saunders College Publishing, Philadelphia, PA.

Carefully insert the rubber stopper into the graduated cylinder. Avoid trapping air, and obtain a good seal. Invert the cylinder into the electrolyte in the beaker and loosen and remove the rubber stopper. The cylinder must remain full. Clamp the cylinder in place so that the 25-mL mark on the cylinder corresponds with the level of electrolyte in the beaker, and rinse your hands.

Lightly sand the Zn electrode, then wipe it off with a Kimwipe. After sanding, handle the electrode with clean forceps or tongs. Find the mass of the electrode to 0.1 mg. Assemble the cell, attaching the electrical wire to the Zn electrode with an alligator clip and using electrical tape to fasten the wires to the battery. Be sure the bare end of the Cu wire is entirely within the graduated cylinder. Start the electrolysis by immersing the Zn electrode in the electrolyte so that the alligator clip does not get wet. You should see bubbles of $H_2(g)$ on the cathode. While waiting for the two electrolyses to finish (about 2 hours), carry out the gas collection experiment.

Direct Reaction. Find the mass of a sample of zinc to 0.1 mg. Be sure that the mass lies in the range 0.100-0.120 g. Then set up the apparatus shown in **Figure 2**. Clamp the syringe horizontally, but do not over-tighten the clamp. BE CAREFUL NOT TO ALLOW THE PLUNGER TO FALL OUT OF THE SYRINGE. Connect the tygon tubing (B) securely to the glass tube (C), but do not connect it to the syringe yet. Rinse the test tube (E) with distilled water and dry with a paper towel. Measure 10 mL of 6M HCl and pour it carefully into the test tube. Then use a paper towel to dry the inside of the test tube above the liquid level. Insert a small wad of glass wool into the test tube and push it down using a stirring rod until it is 1/2 inch above the liquid level. DO NOT ALLOW THE GLASS WOOL TO BECOME WET! The glass wool will keep the metal separated from the acid until you initiate the reaction. Carefully add your weighed metal sample to the tube, so it rests on the glass wool. Insert rubber stopper D into the test tube and seat it firmly. Attach rubber tubing B to the tip of the syringe. Tip the test tube and agitate it gently to initiate reaction. You should see gas form as the reaction proceeds. Continue agitating until reaction is over. There should be no residual metal in the tube. Place the tube in a test tube rack and let it cool to room temperature (2-3 minutes). Lightly twist the syringe plunger to free it from sticking, then record the volume of gas in the syringe to the nearest 0.1 mL. Also record room temperature, the vapor pressure of water at this temperature, and atmospheric pressure. Dispose of materials as indicated below. Using the results from the first run, calculate the mass of Zn needed to produce 45 mL of hydrogen. Measure approximately this amount of metal and repeat the experiment. When finished, disassemble the apparatus and return it to the instructor.

Figure 2
Direct Reaction

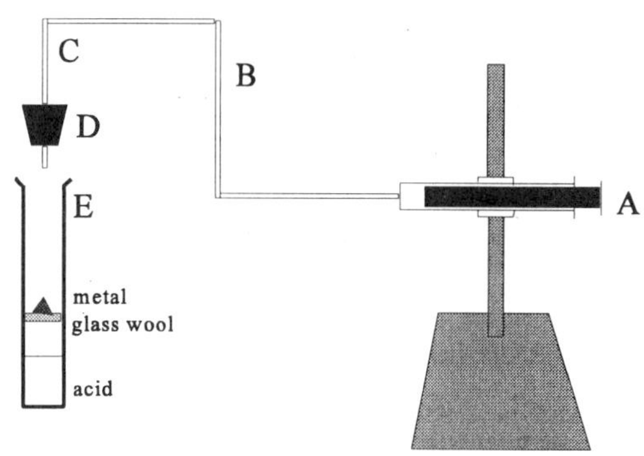

Electrolysis. Remove the Zn electrode from solution when the level of liquid in the cylinder corresponds with the level of liquid in the beaker. Disconnect the dry cell. Rinse the Zn electrode (distilled H_2O), dry it completely with a paper towel, and reweigh it. Record room temperature, atmospheric pressure, and the volume of gas in the graduated cylinder. Calculate the moles of $H_2(g)$ in the cylinder using the ideal gas law, correcting for the partial pressure of water vapor in the cylinder.

When finished, return the Zn electrodes, wires, batteries, beakers, and graduated cylinders.

The instructor will tell you how to submit your data relating volume of hydrogen collected to mass of Zn.

Disposal Methods

Direct Reaction. Remove the stopper from the test tube and pour the solution into the sink. Throw the glass wool plug in the waste basket. Flush the acid solution down the drain.
Electrolysis. Pour the electrolyte down the drain.

Questions

1. Why must the alligator clip be kept dry when the cell is assembled for electrolysis?
2. What effect would each of the following errors have on the results of the electrolysis experiment?
    a. The copper electrode was placed so that some H_2(g) did not go up the graduated cylinder.
    b. The alligator clip was submerged in electrolyte.
    c. The pressure of water vapor in the cylinder is ignored.
    d. The Zn electrode was not dried thoroughly prior to finding the final mass.
    e. The concentration of acetic acid in the electrolyte was less than 0.1M.
3. What effect would each of the following errors have on the atomic weight calculated from direct reaction data?
    a. You used 9 mL of 6M HCl instead of 10.
    b. The rubber stopper was not sealed well in the test tube and some gas leaked out.
    c. You failed to twist the syringe plunger, so it stuck in the barrel at the 25-mL mark.

References

1. Hunt and Block, Laboratory Experiments for General Chemistry, Saunders College Publishing, Philadelphia, 1990.

Preparation
Atomic Weight

Name ____________
Section ____________
Date ____________

Read

1. This experiment
2. The appropriate sections of your textbook.

Problems

1. An electrolysis is carried out at 20 °C and an atmospheric pressure of 689 torr using the apparatus of Figure 1 with a nickel (instead of zinc) electrode. 21.4 mL of H_2(g) are collected during the electrolysis. At the same time, 0.0476 g of nickel are lost from the electrode. Calculate the atomic weight of nickel.

Results
Atomic Weight

Name ____________________
Section ____________________
Coworker ____________________
Date ____________________

Data

Electrolysis

	Run 1	Run 2
mass of zinc electrode		
before electrolysis	_____	_____
after electrolysis	_____	_____
room temperature	_____	_____
atmospheric pressure	_____	_____
vapor pressure of H_2O	_____	_____
volume $H_2(g)$ collected	_____	_____
moles $H_2(g)$ collected	_____	_____
atomic weight of Zn	_____	_____

CALCULATIONS:

Direct Reaction Method

	Run 1	Run 2
mass sample	_____	_____
initial volume	_____	_____
final volume	_____	_____
V_{gas} in syringe	_____	_____

T = _____°C
P = _____torr
$P_{vap}(H_2O)$ = _____ torr

CALCULATIONS:

Answers to Questions

Calorimetry

1 lab period; work in pairs. Complete the Preparation page before laboratory.

Goals

- To accurately measure temperature changes during chemical reaction
- To discover quantitative aspects of heat produced by reactions
- To understand that enthalpy is extensive

Background

Chemical reactions are accompanied by energy changes. Most often, the energy manifests as heat. Heat may be either produced or absorbed, depending on the reaction. A reaction that produces heat is exothermic; one which absorbs heat, endothermic. Most spontaneous reactions (those which occur unaided) are exothermic. If a reaction is carried out in a closed, insulated container, the heat produced or absorbed causes a temperature change (ΔT) of the container contents. Measurement of ΔT allows quantitative determination of the heat produced per mole of chemical reaction. This experimental technique is called calorimetry.

A chemical reaction between reagents A and B in aqueous solution is shown in equation 1. For discussion, we assume the reaction is exothermic--produces heat.

$$A(aq) + B(aq) \rightarrow Products(aq) + heat \qquad \text{Exothermic} \quad \Delta H < 0 \qquad (1)$$

To measure the heat produced, we follow a very careful and precise procedure:

1. Prepare solutions of reactants A and B of precisely known concentration.
2. Allow the temperatures of solutions A and B to stabilize.
3. Place a precisely known volume of reactant A in the calorimeter.
4. Measure and record the temperature of reactant A in the calorimeter at precise intervals for a specific time (≈two minutes). During this time adjust the temperature of reactant B until it is the same as that of reactant A.
5. At time t_o, quickly add a precisely known volume of reactant B to the calorimeter with good stirring.
6. Measure and record the temperature of the solution in the calorimeter at precise intervals for a specific time.
7. Plot the temperature-time data. A typical plot for an exothermic process is in Figure 1.
8. Measure ΔT for the reaction by extrapolating the pre- and post-reaction linear portions of the T-t curve to time t_o.

Hess' Law. Hess' Law states that enthalpy, H, is a state function. The value of ΔH for conversion of a set of reactants to a set of products is independent of how the conversion is done. For example, consider reaction of AB with C. Two different ways of carrying out this reaction are shown below.

Figure 1
Temperature-Time Plot

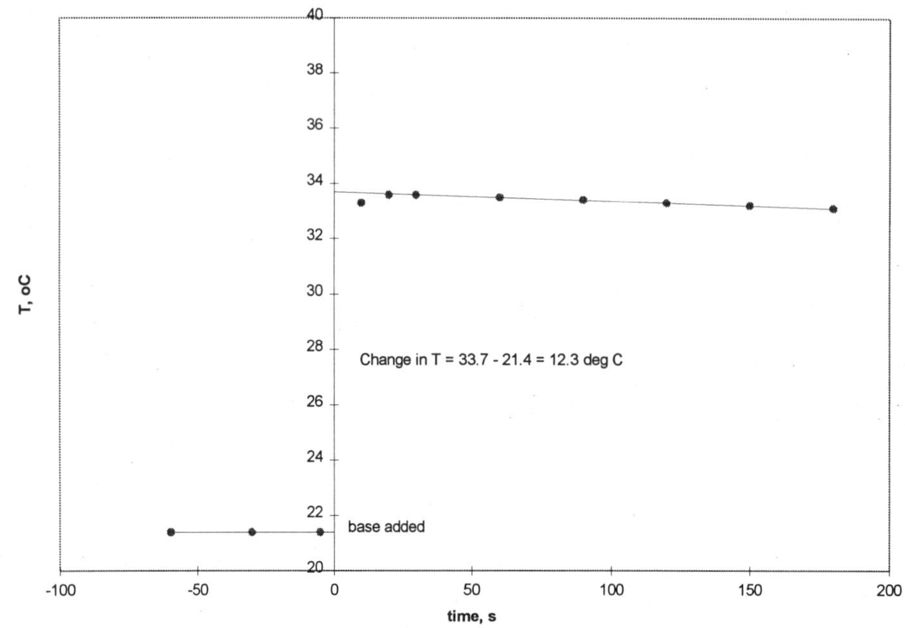

Path 1, direct reaction:

a) $AB + C \rightarrow AC + B \quad \Delta H_a$

Path 2, indirect reaction, involving two steps:

b) $AB \rightarrow A + B \quad \Delta H_b$

c) $A + C \rightarrow AC \quad \Delta H_c$

The sum of reactions b and c is reaction a. Since paths 1 and 2 start and end with the same substances, the total change in enthalpy along path 1 is the same as the total change in enthalpy along path 2.

$$\Delta H_a = \Delta H_b + \Delta H_c$$

If any two of these enthalpies are known or measurable, the third may be readily calculated, even if it is <u>not</u> directly measurable! This is the value of Hess' Law.

Aqueous Solutions of Acids and Bases. In this experiment, we work with two strong acids, HCl and H_2SO_4, and the strong base NaOH. When dissolved in water, these three substances ionize completely, as indicated below.

$$HCl(aq) \rightarrow H^+(aq) + Cl^-(aq) \quad (2)$$

$$H_2SO_4(aq) \rightarrow H^+(aq) + HSO_4^-(aq) \quad (3)$$

$$NaOH(s) \rightarrow Na^+(aq) + OH^-(aq) \quad (4)$$

Thus a 0.1 M solution of HCl actually contains 0.1 mole H^+ and 0.1 mole Cl^- per liter of solution. It contains no unionized HCl molecules. Similarly, a 0.1 M solution of H_2SO_4 contains 0.1 mole H^+ and 0.1 mole HSO_4^- per liter; and a 0.1 M solution of NaOH contains 0.1 mole Na^+ and 0.1 mole OH^- per liter.

When we mix a solution of HCl with a solution of NaOH, a neutralization reaction occurs between H^+ and OH^-:

$$H^+(aq) + OH^-(aq) \rightarrow H_2O \quad \Delta H_5 = -55.835 \text{ kJ/mole} \quad (5)$$

The Cl^- ion from HCl and the Na^+ ion from NaOH undergo no reaction, so are not explicitly written in the neutralization equation. When we mix a solution of H_2SO_4 with a solution of NaOH, the situation is more complex. Two reactions must take place during neutralization of H_2SO_4:

$$H^+(aq) + OH^-(aq) \rightarrow H_2O \quad \Delta H_6 = \Delta H_5 = -55.835 \text{ kJ/mole} \quad (6)$$

$$HSO_4^- + OH^-(aq) \rightarrow H_2O + SO_4^{2-} \quad \Delta H_7 \quad (7)$$

The total enthalpy change for neutralization of a solution of H_2SO_4 by NaOH, ΔH_{total}, is therefore $\Delta H_6 + \Delta H_7$, by Hess' Law.

In this experiment, you will use calorimetry to discover an answer to the following question:

How is ΔH for reaction of a strong acid with NaOH related to the number of acidic hydrogen ions in the acid?

Equipment and Materials

graph paper
thermos calorimeter
bent glass stirring rod
timer
2 thermometers
50-mL burets
50-mL graduated cylinder
100-mL graduated cylinder
standard NaOH, ≈2M
standard HCl, ≈2M
standard H_2SO_4, ≈2M

Safety

Safety glasses must be worn at all times in the laboratory. You will work with solutions of acids and bases. Avoid ingestion and contact of these solutions with the skin. If you spill an acid or base solution on your skin, wash the area immediately with plenty of cold water. BE PARTICULARLY CAREFUL WHEN FILLING THE BURET WITH 2.0 M ACID.

Experimental

Record all data in your lab notebook or on the **Results** sheet, whichever is appropriate. Each team will be assigned 2 or 3 volumes of acid solution to react with 50 mL of 2.0 M NaOH. Obtain about 160 mL of NaOH solution and a volume of acid equal to the sum of your assigned volume + 15 mL. Rinse a 50-mL buret with 2 5-mL portions of your acid, then fill the buret with acid. Measure the ΔT for addition of 50 mL of 2.0 M NaOH to each assigned acid volume in the calorimeter, using the general procedure outlined in the **Background**. For results to be comparable, all teams must have the same total volume of solution--100 mL--in the calorimeter at the end of the reaction. Add a volume of distilled water equal to (50 - assigned acid volume) mL to the calorimeter using a 50-mL graduated cylinder. Then add the assigned volume of acid solution to the calorimeter using a buret. Pour 50 mL NaOH solution into the 100-mL graduated cylinder. Measure the temperature of the solution in the calorimeter and the NaOH solution in the graduate. They should be the same to within 0.2 °C. If necessary, adjust the temperature of the NaOH solution by running hot or cold water over the outside of the graduate. BE CAREFUL NOT TO GET ANY OF THIS WATER INTO THE GRADUATE. After temperature equilibration, measure and record the temperature (to the nearest 0.1°C) of the solution in the calorimeter at t = -60 sec, t = -30 sec, and t = -5 sec. At t = 0, quickly remove the calorimeter lid and pour the 50 mL NaOH solution into the calorimeter from the 100-mL graduated cylinder. Total volume in the calorimeter will be 100 mL. Record temperature at t = 10, 20, and 30 sec and subsequently at 30 second intervals for a total time of 3 minutes. Repeat for the second and third runs. Make all three T-t plots on the same sheet of graph paper, and determine ΔT for each run. Process data as directed by the instructor. Work on **Questions** until the postlab discussion.

Disposal Methods

Pour spent reagents carefully into the sink, and flush down the drain with plenty of water.

Questions

While waiting for discussion, work on the following questions.

1. In Figure 1, why does T stay constant before addition of reagent B?
2. In Figure 1, why does T steadily decrease after reaction of A and B?
3. In Figure 1, why does the full temperature change, ΔT, not take place instantaneously when B is added?
4. Suppose that a plot of ΔT versus volume HCl in the calorimeter was constructed from class data. What would such a plot look like?
5. Suppose you add 50 mL 2 M NaOH solution to 20 mL 2 M HCl solution and measure ΔT_1. Then you add 50 mL 2 M NaOH solution to 20 mL 2 M H_2SO_4 solution and measure ΔT_2. How are ΔT_1 and ΔT_2 related?

References

1. "A Laboratory-Centered Approach to Teaching General Chemistry" by Robert W. Ricci and Mauri A. Ditzler, *Journal of Chemical Education*, **1991**, 68, 228-231

Preparation
Calorimetry

Name ________________

Section ________________

Date ________________

Read

1. This experiment.
2. The appropriate sections of your textbook.

Problems

1. $\Delta H = -55.835$ kJ for the reaction of 1 mole HCl with 1 mole NaOH in a total volume of 1 L of aqueous solution. Suppose you react 50 mL 2 M NaOH with 50 mL 2 M HCl and measure a temperature increase of 12.3°C. What is the heat capacity (in J/°C) of calorimeter + 100 mL of solution?

Results
Calorimetry

Name ________________

Section ________________

Coworkers ________________

____

____

____

Date ________________

Data

Acid ________ Concentration_____ Volume ____mL

NaOH Concentration_____

	Run 1		Run 2		Run 3

Volume acid:

Moles acid:

time	T	time	T	time	T

$\Delta T =$ ______ ______ ______

Answers to Questions

Precipitation†

1 period; work individually. Complete the Preparation page before laboratory.

Goals

- To make thermodynamic predictions of cation/anion reactions in aqueous solution
- To carry out reactions of cations and anions in aqueous solution
- To devise solubility rules

Background

A spontaneous process is one that takes place without assistance from outside. Examples are melting of ice at room temperature, rusting of iron, and combustion of natural gas in a furnace. A process is spontaneous either because it is exothermic (has $\Delta H < 0$) or because it increases disorder (has $\Delta S > 0$) or both. The effects of these two quantities on spontaneity are neatly summarized in terms of a single quantity, called the free energy, which is related to enthalpy (H) and entropy (S) by equation 1.

$$\Delta G = \Delta H - T\Delta S \qquad (1)$$

Processes for which $\Delta G < 0$ are spontaneous; those for which $\Delta G > 0$ are not.

It is useful for chemists to tabulate the standard free energies of formation, ΔG_f^o, for many chemical substances. ΔG_f^o is the change in free energy which accompanies the formation (subscript f) of one mole of a substance in its standard state (superscript ^{o}) from its elements in their standard states. For example, the formation of one mole of $CaCO_3(s)$ from its elements is shown below:

$$Ca(s) + C(s) + 3/2\ O_2(g) \rightarrow CaCO_3(s)$$

The free energy change which accompanies this process is, by definition, the standard free energy of formation of $CaCO_3$. ΔG^o for any chemical reaction may be calculated from the ΔG_f^o values of reactants and products using equation 2. This is what makes ΔG_f^o so useful. To calculate the free energy change for a

$$\Delta G^o \text{ for reaction} = \Sigma\ \Delta G_f^o(\text{products}) - \Delta G_f^o(\text{reactants}) \qquad (2)$$

chemical reaction under conditions other than standard, we need only correct ΔG^o for the non-standard conditions using equation 3.

$$\Delta G = \Delta G^o + RT \ln Q \qquad (3)$$

Here R is the gas constant, in kJ/mole-K, T is the Kelvin temperature, and Q is the mass action quotient.

A reaction of two water soluble salts in which cation and anion partners are "traded" is called a metathesis or double displacement reaction. Such a reaction is represented schematically below, where A and B are two different cations, X and Y two different anions.

---

†Adapted with permission from an experiment developed by Professor CJ Scaife, Union College, Schenectady, NY.

$$AX + BY \rightarrow AY + BX \quad (4)$$

For such a reaction to proceed to the right, there must be a thermodynamic driving force. The driving force is provided, according to Le Chatelier's principle, by removing one of the products from solution. This will occur if one of the products precipitates (i.e., leaves solution as a solid), is a weak electrolyte (i.e., is essentially unionized in solution), or decomposes to form a gas, which escapes from solution. Reaction of two salts to form solid, weak electrolyte, or gas is spontaneous. The free energy change, ΔG, for such a process is negative. This experiment deals with reactions that (potentially) form solids.

The formation of a precipitate is illustrated in equation form below. Three equations, of increasing simplicity, are shown.

$$\underset{\text{soluble}}{AgNO_3(aq)} + \underset{\text{soluble}}{NaCl(aq)} \rightarrow \underset{\text{insoluble}}{\mathbf{AgCl(s)}} + \underset{\text{soluble}}{NaNO_3(aq)} \quad (5)$$

$$Ag^+(aq) + NO_3^-(aq) + Na^+(aq) + Cl^-(aq) \rightarrow \mathbf{AgCl(s)} + Na^+(aq) + NO_3^-(aq) \quad (6)$$

$$Ag^+(aq) + Cl^-(aq) \longrightarrow \mathbf{AgCl(s)} \quad (7)$$

These are called, respectively, the total equation (TE), the ionic equation (IE), and the net ionic equation (NIE). The TE shows what reagents to obtain from the stockroom to carry out a given reaction. An IE is obtained by writing all soluble strong electrolytes from the TE in dissociated form; it bridges the total equation and the net ionic equation. The net ionic equation is obtained by omitting ions that appear on both sides of the ionic equation. Canceled ions are called spectator ions because they do not take part in the chemical reaction. Therefore, the NIE shows in simplest form what species take part in the reaction.

We can readily calculate ΔG for any net ionic equation from ion concentrations and ΔG_f^o values for reactants and products. Table 1 gives standard free energies of formation, ΔG_f^o, for 1 M solutions of several cations (first and second columns) and anions (third and fourth columns). The fifth and sixth columns show ΔG_f^o for several solid salts formed from combinations of the listed cations and anions. Note that two of the solids crystallize with waters of hydration (e.g., $CaSO_4 \bullet 2H_2O$). Equations for formation of these solids must include the appropriate number of water molecules on the left side. ΔG_f^o for water is -237.18 kJ/mole.

Table 1

ΔG_f^o Values in kJ/mol for Cations, Anions, and Solids

(Data from the CRC Handbook of Chemistry and Physics, 63rd Edition, CRC Press, Inc., Boca Raton, FL, 1982)

Cations	ΔG_f^o	Anions	ΔG_f^o	Selected Solids	ΔG_f^o
Ag^+	77.12	Cl-	-131.26	AgCl	-109.80
Ba^{2+}	-560.74	I-	-51.59	Ag_2SO_4	-618.48
Ca^{2+}	-553.54	SO_4^{2-}	-744.63	AgI	-66.19
Pb^{2+}	-24.39	CrO_4^{2-}	-727.85	$BaCrO_4$	-1345.3
Al^{3+}	-485.34	OH-	-157.29	$CaCl_2$	-748.1
				$Pb(OH)_2$	-452.29
				$BaCl_2 \bullet 2H_2O$	-1296.5
				$Al_2(SO_4)_3$	-3100.1
				$PbSO_4$	-813.2
				$CaSO_4 \bullet 2H_2O$	-1797.4

Using data from the table, we calculate ΔG^o for equation 7 as follows:

$$\Delta G^o = \Delta G_f^o(AgCl(s)) - \Delta G_f^o(Ag^+(aq)) - \Delta G_f^o(Cl^-(aq))$$

$$= -109.789 - 77.107 - (-131.228) \text{ kJ/mole}$$

$$= -55.668 \text{ kJ/mole}$$

This number is substantially negative. Formation of solid AgCl(s) from its ions under standard conditions (i.e., $[Ag^+] = [Cl^-] = 1$ M) is spontaneous--it will occur of its own accord. If ion concentrations were 0.1 M rather than 1 M, would precipitation still occur? Use equation 3:

$$\Delta G = \Delta G^o + RT \ln Q$$

$$\Delta G = \Delta G^o + RT \ln 1/[Ag^+][Cl^-]$$

$$= -55.668 + (8.314 \times 10^{-3})(298) \ln [1/(0.1)(0.1)]$$

$$= -44.250 \text{ kJ/mole}$$

Precipitation is still spontaneous, but somewhat less so than at 1 M ion concentrations.

Solubility is the amount of solute (in g or moles) that dissolves in a given amount of solvent (in mL, L, or g). If quite a bit of a solid dissolves in a solvent, the solid is soluble in that solvent. If only a small amount of a solid dissolves in the solvent, the solid is insoluble, or sparingly soluble, in the solvent. A solubility of 1.0 gram solute per 100 mL solvent is considered the border line between soluble and insoluble. For example, 74.5 g of calcium chloride, $CaCl_2$, dissolves in 100 mL H_2O; thus $CaCl_2$ is soluble in water. In contrast, only 2.9×10^{-4} g silver chloride, AgCl, dissolves in 100 mL water, so it is insoluble in water. Formation of a soluble salt from its ions has $\Delta G^o > 0$; formation of an insoluble salt from its ions has $\Delta G^o < 0$.

In this experiment, you investigate many cation-anion combinations to discover which produce insoluble products. First, before coming to lab, predict whether or not solid will form when specified cation/anion pairs are brought together. Then, the cation/anion combinations that you predicted, and many others as well, will be investigated in the laboratory to learn about double displacement reactions which form solids. To minimize time, cost, and safety hazard, and to maximize the efficiency of carrying out and cleaning up the experiments, you will work on a very small scale (microscale). Based on observation, you will propose answers to the following questions:

Can your observations about solubility be summarized in the form of a few rules?
Do theoretical predictions of solubility correlate with observation?

Equipment and Materials

8.5 x 11" plastic reaction sheet
cation/anion grid
paper clips
disposable micropipets
magnifying lens

Kimwipes
microstirrers (toothpicks)
wash bottle
red litmus
6 M NaOH
1 M Na_2S
solutions containing **cations (+ nitrate anion)** and **anions (+ sodium cation)**

Safety

All reagents used in this experiment are to be considered harmful. Wear your goggles. Avoid ingestion of solutions or solids.

Experimental

Record all data and observations in your lab notebook or on the **Results** sheet, whichever is applicable. Obtain a plastic reaction sheet, a paper grid, two paper clips, and three microstirrers (toothpicks). Wipe the sheet with a Kimwipe to remove any residue (contaminants may cause confusing reactions). Attach the grid to the underside of the plastic sheet with paper clips. Practice delivering individual drops of water from a disposable micropipet until you can do it competently.

Several test tube racks, each containing test tubes with cation and anion solutions, and micropipets for dispensing solution, have been placed in the lab. You will share the solutions and pipets with others, so be quick, careful, and courteous. Do not keep a test tube/micropipet set at your bench for longer than you need it. Start with the cation nitrate solutions shown in the left column of the grid. The cation nitrate solutions were prepared by dissolving the nitrate salt of the cation in water (e.g., $Ca(NO_3)_2(s)$ was used for Ca^{2+}). Obtain a test tube containing a solution of any cation except Na^+ (which will be omitted, since it is already present in all the anion solutions). Use the pipet to place one drop of the cation solution in the appropriate cation circle in the leftmost column on the grid. Place nine more drops in the smaller circles horizontally across the grid. Center the drops in the circles. Return the pipet to the test tube, and the test tube to the rack. Continue with other cation solutions until you have a complete matrix of cation drops (160 total).

Proceed similarly with the sodium anion solutions in the top row of the grid. Sodium anion solutions were prepared by dissolving the sodium salt of the anion in water (e.g., NaBr(s) was used for Br^-). Obtain a test tube containing solution of any anion except NO_3^- (omitted, since it is already present in all cation solutions). Use the pipet to place one drop of the anion solution in the appropriate anion circle in the top row on the grid. Place 16 more drops vertically down the sheet. Place each drop directly on the drop of cation solution. Complete the matrix of anion solutions (153 drops in all). Use only one drop of each solution, and be careful not to touch the tip of the dropper to other solutions on the sheet. Finally, place one drop $NaNO_3(aq)$ centered in the oval in the upper left corner of the grid.

Record observations on the boxed grid (see following pages). Indicate whether solid is formed, and colors of solids and/or remaining solutions. Cloudiness often suggests solid formation. If in doubt about solid formation, view the drop with a magnifying lens and/or stir the drop carefully with a microstirrer. Use a colored highlighter to indicate on the boxed grid where precipitate formed.

Disposal Methods

Clean the plastic reaction sheet by curling it and emptying the solutions and solids into a 250-mL beaker.

Rinse remaining material into the beaker using a distilled water wash bottle. USE NO MORE THAN A TOTAL OF 100 mL OF WATER TO WASH THE SHEET. Finally, rinse the working side of the sheet with tap water and dry carefully with a towel, then with a Kimwipe. When the sheet is completely dry, return it and the paper grid sheet to the instructor.

UNDER A FUME HOOD, add 6 M NaOH solution dropwise to the 250-mL beaker until the mixture is basic to litmus (3-5 drops required). Add 1 drop excess. Add 10 drops 1 M Na_2S solution to insure complete precipitation of aqueous cations as insoluble sulfide salts. Finally, heat the mixture using a Bunsen burner just to boiling for 3 minutes to coagulate the precipitate. Pour the mixture into the waste bottle in the hood labelled HEAVY METAL MIXTURE. This will later be filtered to remove heavy metal waste prior to disposal. Wash your hands well.

Questions

1. How accurate were your thermodynamic predictions? Prepare a table showing predicted and observed results for each cation/anion combination.
2. What cation(s) form only soluble salts with the anions investigated? What cation(s) form only insoluble salts with the anions? What anions form only soluble salts with the cations investigated? What anions form only insoluble salts with the cations?
3. Formulate a set of solubility rules to summarize your experimental observations for the anions. For anions that form primarily soluble salts, a rule might take the form: all chlorides are soluble except... For anions that form primarily insoluble salts, a rule might take the form: all hydroxides are insoluble except... Group several anions in a single rule when possible.

References

1. C. Scaife, Union College, communication.

Preparation
Precipitation

Name _______________
Section _______________
Date _______________

Read

1. This experiment.
2. The appropriate sections of your textbook.

PRIOR TO LAB. Use the data in Table 1 to predict whether or not the solids listed in the table will form when a 0.2 M solution of cation is mixed with an equal volume of a 0.2 M solution of anion (allow for a two-fold dilution of each ion due to mixing!). In all you should make ten predictions before coming to lab.

Problems

1. Write total equation, ionic equation, and net ionic equation for the reaction of Ag^+ with each anion.

Results
Precipitation

Name ______________
Section ______________
Date ______________

Thermodynamic Predictions

Salt	ΔG	Soluble?

Precipitate = yes or no
Y n

Observations

Name ______________

Section ______________

Date ______________

		1	2	3	4	5	6	7	8	9
	NO_3^-	Cl^-	Br^-	I^-	CO_3^{2-}	SO_4^{2-}	PO_4^{3-}	CrO_4^{2-}	OH^-	S^{2-}
Na^+	X	x	x	x	x	x	x	x	x	x
A K^+	X	n	n	n	n	n	n	n	n	n
B NH_4^+	X	n	n	n	n	n	n	n	n	n
C Mg^{2+}	X	n	n	n	y	n	y	n		
D Ca^{2+}	X	n	n	n	y	n	y	n	y	y
E Ba^{2+}	X	n	n	n	y	y	y	y	n	y
F Al^{3+}	X	n	n	n	y	n	y	y	n	n
G Pb^{2+}	X	n	y	y	y	y	y	y	y	y
H Cr^{3+}	X	n	n	n	y	n	y	y	y	n
A Mn^{2+}	X	n	n	n	y	n	y	y	y	y
B Fe^{2+}	X	n	n	n	y	n	y	y	y	y
C Co^{2+}	X	n	n	n	y	n	n	n	y	n
D Ni^{2+}	X	n	n	n	y	n	y	n	y	y
F Cu^{2+}	X	n	n	y	y	n	y	y	n	n
G Zn^{2+} (Bottle)	X	n	n	n	y	n	n	y	n	y
H Ag^+	X	y	y	y	n	n	y	y	n	y
A Hg_2^{2+}	X	y	y	y	y	y	y	y	y	y

Reactions of Copper

1 lab period; work individually. Complete the Preparation page before laboratory.

Goals

- To learn synthetic techniques
- To determine the yield of a series of chemical transformations
- To identify the type of reaction, reactants, products, and limiting reagent for each step
- To write balanced chemical equations for each reaction

Background

When new substances are synthesized, several factors must be considered. If reactants are ions that form insoluble products, then the products can be isolated by filtration. In this simple type of metathesis, one reagent is in excess and the excess must be enough to drive the reaction to completion by Le Chatelier's principle. Other reactions involve the transfer of electrons, redox reactions. In this experiment, you will be using both types of chemical transformations to study reactions of Cu.

Copper is a transition metal capable of forming colored compounds. For example, malachite is a green mineral $Cu_2(CO_3)(OH)_2$ used for making ornamental objects and jewelry. Similarly, azurite is a blue semiprecious stone and is also a carbonate of copper, $Cu_3(OH)_2(CO_3)_2$. Blue $CuSO_4 \cdot 5H_2O$ is used as a disinfectant.

What factors determine the product yield in a synthesis?

Equipment and Materials

Buchner funnel
filter flask
10-mL graduated cylinder
100-mL graduated cylinder
glass stirring rod
Cu turnings
concentrated HNO_3 (15.9 M)
concentrated H_2SO_4 (12 M)
Zn metal
saturated $NaHCO_3$

Safety

Eye protection must always be worn in the lab. Acids used in this experiment are very concentrated. Wear gloves when handling them. Wash any spills with soap and water. Fumes from the reaction of Cu with nitric acid are toxic. Add the acid only under a fume hood.

Experimental

Record your data in your notebook or on the **Results** page as instructed. Write a balanced chemical equation for each step as you proceed. Weigh a ball of Cu turnings by difference to 0.1 mg. Prepare a dilution of HNO_3 by adding 5 mL of concentrated acid to 8 mL of H_2O. Under the HOOD add the diluted acid to the Cu turnings. Allow the Cu to completely dissolve. Slowly add the Cu solution to 75 mL of saturated $NaHCO_3$ with vigorous stirring. Do not transfer any remaining solids; rinse these with 1-2 mL of water and add the rinsings to the bicarbonate solution. Filter through a Buchner funnel and wash with 2-5 mL water. Transfer the dry solid to a beaker with a minimum amount of water. Slowly add 5 mL 12 M H_2SO_4. Add water to a total volume of 75 mL and cool in an ice bath. Add approximately 1.2 g of granular Zn stirring continuously. After this reaction, if the solution is blue, add more Zn; if Zn remains, add more sulfuric acid. Isolate the solid by filtration, and wash with water. Aspirate to dry. Weigh the product. Store the product until the next lab meeting. Weigh the product again at the beginning of the next lab period. Calculate the overall percent yield for these reactions.

Disposal Methods

Place all solutions in the appropriately marked waste bottles. After weighing the product, place it in the appropriate waste bottle.

Questions

1. Write balanced chemical equations and net ionic equations to describe the reactions occurring with the addition of each reagent.
2. What is the limiting reagent in each step? What type of reaction is occurring, metathesis or redox?
3. Report the overall percent yield of these reactions.
4. At what steps was product lost? How could this be improved to increase the yield?

1st balanced equation.

$Cu_{(s)} + HNO_3 \rightarrow Cu^{+2}\ NO_{(g)}$

$3(Cu_s \rightarrow Cu^{+2} + 2e^-)$

$2(4H^+ + 3e^- + NO_3^-{}_{(aq)} \rightarrow NO_{(g)}\uparrow + 2H_2O)_3$

$3Cu \rightarrow Cu^{2+} + 6e^-$

$8H^+ + 6e^- + 2NO_3^- \rightarrow 2NO_g\uparrow + 4H_2O$

$8H^+ + 3Cu + 2NO_3^- \rightarrow 3Cu^{+2} + NO\uparrow + 4H_2O$

›reparation
₹eactions of Copper

ɬame ______________

ʖection ______________

)ate ______________

₹ead

. This experiment.

. About copper in your text.

. **Determining Mass** in this manual

ʅuestions

›efine the following terms:

oxidation

reduction

limiting reagent

percent yield

'ractice Problems

. Balance the following reaction:

$$Cu + HNO_3 \rightarrow Cu^{2+} + NO(g)$$

. In the procedure, nitric acid is diluted. What is the new molarity of the nitric acid used? Concentrated nitric cid is 15.9 M.

36.156 – glass

30.366

Weigh watch glass: 20.027g.
weigh w/ 1g copper: 21.037g.
copper = 1.01g.

Results
Reactions of Copper

ame _____________
ection _____________
ate _____________

HNO_3 sol = 8mL H_2O + 5mL conc. acid

Then add copper to solution

---

ata (be very precise)

Filter = poor blue liquid on ~~sol~~ filter slowley = leaves only solid
I rinsed out my beaker. then put solid blue in clean beaker.
(my blue

issolution of Cu

Mass of container + Cu ______

Mass of container ______

Mass of Cu used ______

Molarity of nitric acid ______

Reaction

Limiting reagent _______ (show calculations)

eaction of copper and sodium hydrogen carbonate
Reaction
Limiting reagent _______ (show calculations)

eaction with sulfuric acid
Reaction
Limiting reagent _______ (show calculations)

eaction with zinc
Reaction
Limiting reagent _______ (show calculations)

lass of product _______

ercent recovery _______

nswers to Questions

Synthesis of $Ni(NH_3)_6Cl_2$

1 lab period; work individually. Complete the Preparation page before laboratory.

Goals

- To learn laboratory techniques
- To prepare 4.0 g of $Ni(NH_3)_6Cl_2$
- To learn about limiting reagents

Background

Frequently, chemists want to prepare a given quantity of a compound and will base the synthesis on information available in the literature. Industrial chemists frequently find the situation where a pilot synthesis must be scaled up to production levels. Many factors influence how a pilot process is scaled. For example, an exothermic reaction may need to be controlled carefully when the amounts of reactants are increased from grams to tons.

In this experiment, you will be scaling down a known synthesis to prepare a smaller amount than that presented in a literature preparation. You will be synthesizing the hexaammine complex of Ni^{2+}. Elemental nickel is used in coinage and in rechargeable batteries where it is used as the cathode. An alloy of nickel with chromium is used in flight recorders because of its resistance to burning in the presence of jet-fuel fires. In aqueous solutions, Ni is most frequently found in the 2+ oxidation state and can form both planar and octahedral complexes.

What is the role of a limiting reagent in a synthesis?
What factors affect the percent yield?
How do you decide if a reaction has occurred?

Equipment and Materials

filter flask
Bunsen burner
Buchner funnel
sample bottle
glass stirring rod
analytical balance
solid $Ni(H_2O)_6Cl_2$
concentrated aqueous ammonia
ethanol (100%)

Safety

Eye protection must always be worn in the laboratory. Use concentrated ammonia in the fume hood.

Experimental

Record your data and observations in your notebook or on the **Results** page as instructed. Plan ahead; if hot or cold solutions are needed, prepare them early.

Dissolve the hydrated nickel(II) chloride in warm water. Add concentrated aqueous ammonia (15 M). Warm the solution for 10 minutes, but do not boil. Cool the solution in an ice bath while stirring constantly. Add the ethanol. Filter and wash the precipitate with cold concentrated aqueous ammonia. Wash the product with alcohol and dry. Determine the mass of the dried product.

Disposal Methods

Turn the product in to the instructor in a bottle labeled with your name, section, and date. All wash solutions should be placed in the appropriately labeled waste bottles unless otherwise directed by your instructor.

Questions

1. What is the mass of the product collected?
2. Based upon the amount of hydrated nickel with which you started, what mass of product was expected?
3. What is the percent yield of your synthesis?
4. What steps, if any, could be taken to improve the percent yield?
5. Do you anticipate any problems scaling up this synthesis?

References

Adapted from Experiments in General Chemistry, 3rd edition, Russell S. Drago, Theodore L. Brown, Allyn and Bacon, Inc., 1970.

Preparation
Synthesis of $Ni(NH_3)_6Cl_2$

Name ____________
Section____________
Date ____________

Read

1. This experiment.
2. About limiting reagents and percent yield in your text.
3. **Determining Mass** in this manual.

Practice Problem

Balance the following equation.

$$Ni(H_2O)_6Cl_2(aq) + NH_3(aq) \rightarrow Ni(NH_3)_6Cl_2(s) + H_2O$$

The above equation is the basis of the experiment outlined below. Determine what the limiting reagent is. Assuming 100% yield, calculate the grams of product expected given the amounts used below.

> Dissolve 32 g of hydrated nickel (II) chloride in 40 mL of warm water. Add 100 mL of concentrated aqueous ammonia (15 M). Warm the solution for 10 minutes, but do not boil. Cool the solution in an ice bath while stirring constantly. Add 50 mL of ethanol. Filter and wash the precipitate with a few mL of cold concentrate aqueous ammonia. Wash the product with a few mL of alcohol, then with acetone and dry.

Limiting reagent ____________

Mass of product based on above synthesis ____________

Calculate the amounts of starting materials needed to prepare 4.000 g of product based upon 100% yield. These will be the amounts you will use in lab.

Reagent	Amount Needed Based on the Synthesis Above	Amount Needed To Prepare 4.0 g of Product
$Ni(H_2O)_6Cl_2$	____________	____________
water	____________	____________
aqueous ammonia	____________	____________
ethanol	____________	____________

Results
Synthesis of $Ni(NH_3)_6Cl_2$

Name ______________
Section______________
Date ______________

Data

Mass of beaker + $Ni(H_2O)_6Cl_2$ __________
Mass of beaker __________
Mass of $Ni(H_2O)_6Cl_2$ used __________

Theoretical yield of $Ni(NH_3)_6Cl_2$ __________
Actual yield of $Ni(NH_3)_6Cl_2$ __________
Percent yield __________

Description of starting material:

Description of the reactions:

Description of product:

Answers to Questions

Synthesis of Hydrated Copper(II) Sulfate

1 lab period; work individually. Complete the Preparation page before laboratory.

Goals

- To prepare 4.000 g of $CuSO_4{\bullet}5H_2O$
- To calculate the amounts of materials needed

Background

Frequently, chemists want to prepare a given mass of a compound and will base the synthesis on information available in the literature. In this experiment, you will be scaling down a known synthesis to prepare a much smaller amount. Industrial chemists frequently find the opposite situation where a pilot synthesis must be scaled up to production levels. The copper compound you prepare is the active ingredient in TidyBowl. Copper is an effective algicide that is also used in the treatment of swimming pools.

How does a chemist scale a synthesis?
What factors must you consider in your synthesis?

Equipment and Materials

filter flask
Buchner funnel
glass wool
Bunsen burner
funnel
analytical balance
25-mL graduated cylinder
glass stirring rod
Cu turnings
15 M HNO_3
18 M H_2SO_4

Safety

Eye protection must always be worn in the laboratory. The acids are concentrated, so wear gloves. Add acid to water. Fumes generated when Cu is mixed with nitric acid are toxic, so add the acid to the Cu under the fume hood.

Experimental

Record your observations and procedures in your notebook or on the **Results** page, as directed. Work at a steady pace. Calculations should be complete so the mass and volumes of reagents needed are known. Prepare ahead. If you will need a cold or hot solution, begin cooling or heating early in the lab period. Carefully record observations and procedures.

Weigh Cu to the nearest 0.1 mg and place in a beaker. Prepare dilute nitric acid by adding concentrated (15.9 M) nitric acid to water. UNDER A FUME HOOD, slowly add the diluted acid to the Cu. Add more nitric acid if the brown fumes cease before all of the Cu is dissolved. When the reaction is complete, add concentrated sulfuric acid (18 M). If any solid remains, filter the solution through glass wool. Evaporate the solvent until crystals begin to form. Place the beaker in cold water until the solution is cooled. Decant the solution and dispose of it as directed below. Add about 2/3 as much water to the solid as the volume of solution decanted. Stir to dissolve the solid. Heat the solution to boiling and set the beaker aside to cool. Filter and dry the solid. Find the mass of product collected.

Disposal Methods

Throw glass wool in the trash. Place the decanted solution in the waste bottle labelled "copper waste." Solid $CuSO_4$ is also placed in the copper waste bottle.

Questions

1. Calculate the theoretical yield based on the amount of Cu used.
2. Calculate the percent yield for the reaction.
3. What factors would be important for scaling up this reaction?
4. What factors caused the yield to be less than 100 %?
5. Are there any changes that could be made to improve the yield?

References

1. Adapted form Experiments in General Chemistry, 3rd edition, Russell S. Drago, Theodore L. Brown, Allyn and Bacon, Inc., 1970.

Preparation
Synthesis of Hydrated Copper(II) Sulfate

Name ____________
Section____________
Date ____________

Read

1. This experiment.
2. **Determining Mass** in this manual.
3. About limiting reagents in your text.

Balance the equations for the synthesis of $CuSO_4{\bullet}5H_2O$.

$$Cu + HNO_3 \rightarrow Cu^{2+} + NO(g)$$

$$Cu^{2+} + H_2SO_4 \rightarrow CuSO_4{\bullet}5H_2O$$

The above equations are the basis of the following synthesis. Assuming 90% yield, calculate the grams of product expected based on the experiment outlined below. What is the limiting reagent? Calculate the amounts of starting materials needed to prepare 4.000 g of product. These will be the amounts used in the laboratory.

> Weigh 5 g of Cu to the nearest 0.1 mg and place in a beaker. Prepare dilute nitric acid by adding 30 ml concentrated (15.9 M) nitric acid to 50 ml of water. **UNDER A FUME HOOD**, slowly add the diluted acid to the Cu. Add more nitric acid if the brown fumes cease before all of the Cu is dissolved. When the reaction is complete, add 15 ml of concentrated sulfuric acid (18 M). If any solid remains, filter the solution through glass wool. Evaporate the solvent until crystals begin to form. Place the beaker in cold water until the solution is cooled. Decant the solution and add about 2/3 as much water as the volume of solution decanted to the remaining solid. Stir to dissolve the solid. Heat the solution to boiling and set the beaker aside to cool. Filter and dry the solid. Weigh the product.

Reagents for synthesis	As presented above	Scaled down for use in lab
Mass of Cu	______	______
Volume of 15 M HNO_3	______	______
Volume of H_2O	______	______
New molarity of HNO_3	______	______
Volume of 18 M H_2SO_4	______	______
Theoretical mass of $CuSO_4{\bullet}5H_2O$	______	______

Results
Synthesis of Hydrated Copper(II) Sulfate

Name ______________
Section ______________
Date ______________

Data

Mass of Cu + beaker ______
Mass of beaker ______
Mass of Cu ______

Volume of 15 M HNO_3 ______
Volume of H_2O ______

New molarity of HNO_3 ______

Volume of 18 M H_2SO_4 ______

Mass of $CuSO_4•5H_2O$ ______

Theoretical mass of
$CuSO_4•5H_2O$ ______

Percent yield ______

Observations:

Answers to Questions

Kinetics of Physical Processes

1 period; work in pairs. Complete the Preparation page before laboratory.

Goals

- To study a simple physical process
- To discover how the rate of the process depends on properties of the system
- To use a spreadsheet to manipulate kinetics data for analysis and presentation

Background

Kinetics is the area of science concerned with the rates of physical and chemical processes--how fast they occur. Although the chemist is concerned primarily with the rates of chemical reactions, the principles involved apply to many common processes.

The rate of a process is determined by measuring the time variation in the value of a property of the system which changes as a result of the process. For example, suppose that the distance that you travel in your car from some starting point varies with time as in Table 1.

Table 1
Distance Traveled as a Function of Time

Distance	Time
0	0
55	1
80	2
140	3
140	4
190	5
240	6
265	7
320	8

Figure 1 shows a plot of distance as a function of time. We can use the figure to discuss two types of rates that scientists frequently work with. The first is called the average rate. The average rate over an interval of time is obtained by dividing the distance traveled in the interval by the time interval.

$$\text{Average rate} = \Delta x/\Delta t \qquad (1)$$

where x represents distance and t, time. For example, over the first hour, the car travels an average of 55 miles/1 hour = 55 miles per hour. Over the second hour, the average rate is 25 miles per hour. Over the first two hours, the average rate is 40 miles per hour. The average rate may be obtained from the graph in Figure 1 as the slope of a straight line drawn between the point at the beginning of the time interval and the point at the end. This is illustrated for the second hour in the figure.

The second type of rate is called the instantaneous rate. This is what you read from the speedometer.

A speedometer displays the rate of travel at the instant that you look at it. The instantaneous rate is obtained from the graph in Figure 1 by measuring the slope of the curve at a particular time. For example, the instantaneous rate at t = 7 hours is illustrated in the figure. The value is 40 miles per hour. The instantaneous rate is an average rate over a very small (infinitesimal) time interval. It is therefore written as in equation 2.

$$\text{Instantaneous rate} = dx/dt \qquad (2)$$

where the symbol d stands for "a very small change in." The rate of travel of the car varies in an irregular manner with time because the distance traveled depends on so many variables (e.g., speed limit, hunger, road conditions, gas mileage, weather, driver comfort, etc.). For this reason, the plot in Figure 1 does not have a simple mathematical form. However, some physical processes have properties that vary in a regular way with time. An example is the rate at which a hot body cools. The temperature, T, of a glass of hot water as it cools varies with time as in Figure 2. The figure reveals that the rate of cooling, -dT/dt (the negative sign is needed to give a positive rate from a negative temperature change), is quite large when the water is much hotter than room temperature. However, the rate smoothly decreases as the water cools, approaching zero as the temperature of the water approaches that of the room. Mathematically,

$$-dT_{water}/dt = k\,(T_{water} - T_{room}) \qquad (3)$$

Equation 3 is called Newton's Law of Cooling. It is an example of a <u>differential rate law</u>, which expresses the rate of change of a property as a function of the value of the property (in this case, temperature). Equation 3 is a first order differential rate law because the rate of temperature change is proportional to the first power of T. The proportionality constant, k, is called the <u>rate constant</u> for the process. It is possible to integrate equation 3 by rearranging the equation so that temperature, T, is on one side and time, t, is on the other:

$$-dT_{water}/(T_{water} - T_{room}) = k\,dt \qquad (4)$$

Integration then gives

$$T_{water} = T_{room} + (T^{o}_{water} - T_{room})e^{-kt} \qquad (5)$$

where T^{o}_{water} is the initial water temperature. Equation 5 is called an <u>integrated rate law</u>. It shows directly how the value of the property (T) depends on time. Many chemical reactions show a regular variation of the reaction rate with the concentrations (or pressures) of the reacting species. Many follow first order differential equations analogous to equation 3. A specific example is nuclear decay, in which the rate of decay is directly proportional to the number of radioactive nuclei present.

In this experiment, the principles and the mathematics of kinetics are applied to simple water flow processes, using common laboratory apparatus. The rate of water flow is measured as the volume of water delivered by the apparatus as a function of time. Your goal will be to answer the following questions:

What is the differential rate law for your water flow process?
What is the integrated rate law for your water flow process?

Figure 1
Plot of Distance versus Time

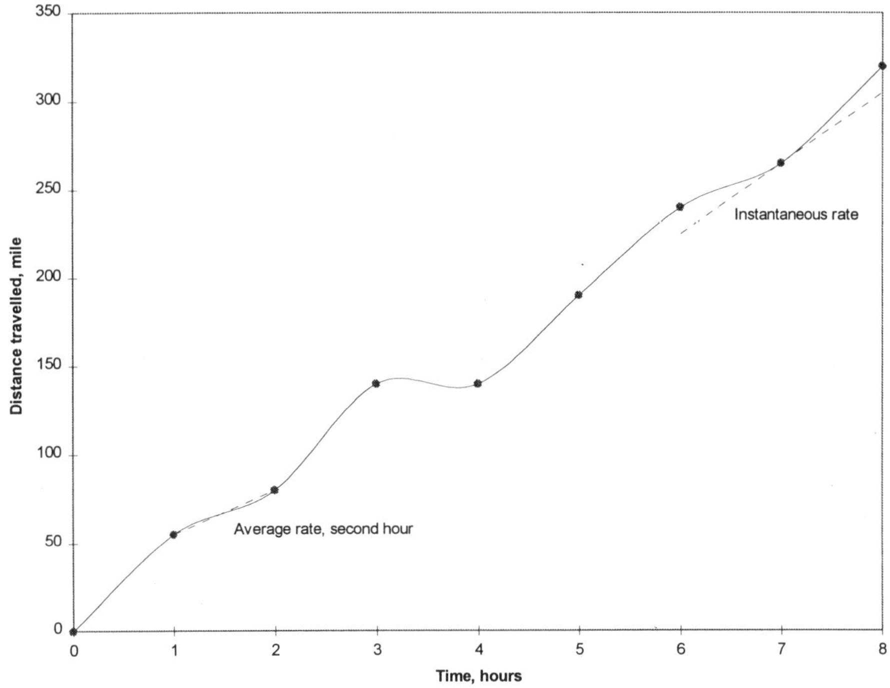

Figure 2
Rate of Cooling

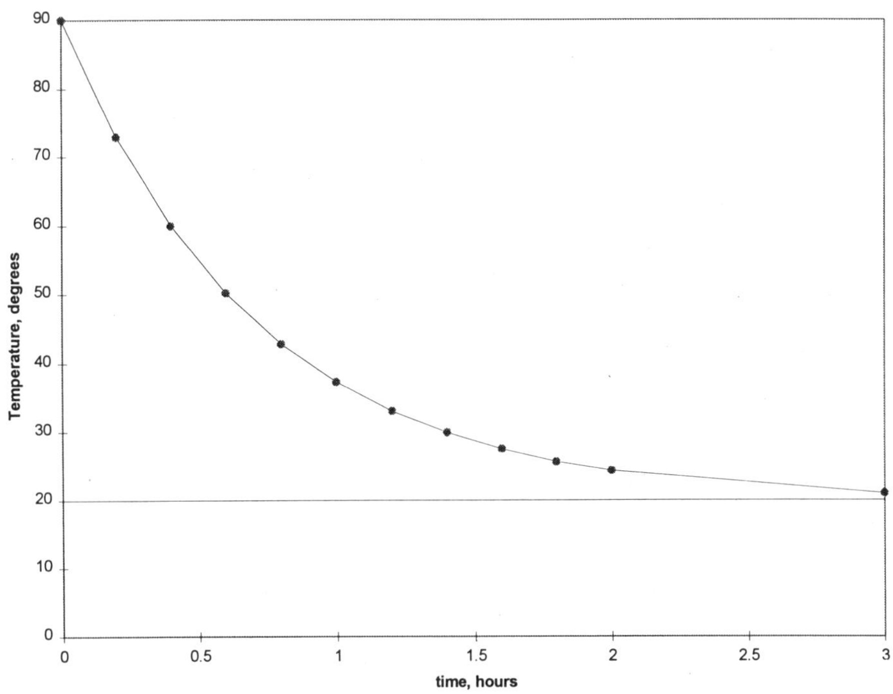

Equipment and Materials

capillary water flow apparatus
timer
50- or 100-mL graduated cylinder
spreadsheet program

Safety

There are no safety concerns with this experiment. However, you must still wear your safety glasses when in the laboratory.

Experimental

Record all data in your lab notebook or on the **Results** sheet, whichever is applicable. Water flow apparatus having a variety of designs are available in the laboratory. Your instructor will assign an apparatus to your team. Instructions for working with the water flow apparatus will be given in class.

Fill up the apparatus to an appropriate level with a known total volume of water. Eliminate all air bubbles from the flow system. Be sure when doing this to collect all water and return it to the apparatus. The capillary should be mounted horizontally. When you are ready, start the timer (you can use the wall clock) and the water flow. Collect the water in a graduated cylinder (50 or 100-mL). At appropriate intervals, note the time and the volume of water collected in the cylinder. Continue the experiment until the water flow has essentially stopped. Replace the capillary tube with one having a different bore size, and repeat the experiment.

Plot the volume-time data for each of your experiments. With the aid of a spreadsheet, determine the differential rate law, which will have the general form

$$-dV/dt = kV^n \qquad (6)$$

Once you have the differential rate law, determine the integrated rate law which will have the general form

$$V = f(V_o, t) \qquad (7)$$

where V_o is the initial volume of water in the apparatus.

If time permits, you may want to carry out experiments with one of the 2-step devices available. How you carry out the experiments is up to you.

Disposal Methods

No disposal required.

Questions

1. On the same set of axes, plot V versus t for the water flow process with two capillary sizes
2. Use a spreadsheet program to determine the differential rate law and k for each of your kinetics runs.
3. Obtain the integrated rate law for your apparatus.
4. The rate of water flow is directly proportional to the height of water in the apparatus, independent of the

shape of the apparatus. Why?
5. Rationalize your differential rate law in terms of the statement in question 4.
6. What is the dependence of the rate constant on the bore diameter of the capillary?

References

1. "Capillary Flow: A Versatile Analog for Chemical Kinetics" by Derek A. Davenport. *Journal of Chemical Education*, **1975**, *52*, 379-381.

Preparation
Kinetics of Physical Processes

Name ________________
Section ________________
Date ________________

Read

1. This experiment.
2. The workshops, **Graphing** and **Spreadsheets**, in this manual
2. The appropriate sections of your textbook.

Problems

1. A hot cup of coffee cools to room temperature, 20°C, as follows. Determine the differential and integral rate laws for the process.

t:	0	0.2	0.4	0.6	0.8	1	1.2	1.4	1.6	1.8	2
T:	90	72.9	60.0	50.2	42.8	37.3	33.0	29.9	27.5	25.6	24.3

2. A sample of 0.0050 g of uranium-238 decays at a rate of 70 disintegrations per second. Determine the rate constant for decay of ^{238}U.

Results
Kinetics of Physical Processes

Name ________________
Coworker ________________
Section ________________
Date ________________

Data

Assigned apparatus (buret, triangular prism, funnel) ____________________

Volume-time data:

time	Volume of water collected

Differential rate law:

Rate Constant: Capillary 1 __________________ Capillary 2 ________________

Integrated rate law

Answers to Questions

Instrumental Experiments

Mixtures

1 lab period; work in groups. Complete the Preparation page before laboratory.

Goals

- To learn the operation of a gas chromatograph
- To learn microliter syringe injection technique
- To discover the use of GC in analysis of mixtures
- To discover the active ingredients in commercial fingernail polish removers

Background

In gas chromatography (GC), a mixture of volatile substances is sent through a long column packed with a stationary material. The column attracts each substance to a different extent, depending upon the structure of the substance. The more strongly a substance is attracted to the column packing, the longer it takes to pass through the column. The total time required for a substance to pass through the GC column and arrive at a detector is called the retention time for the substance. Because the retention time is characteristic of a substance, it may be used to identify the components of a mixture. The intensity of the signal for a pure substance is related to the number of moles of the substance passing through the GC detector. In this experiment, you will use gas chromatography to address the following question:

What is(are) the active ingredient(s) in commercial fingernail polish removers?
What is the approximate mole fraction of active ingredient?

Equipment and Materials

gas chromatograph
injection syringes (1- or 10-µL, depending on GC detector)
pure organic liquids: acetone, ethyl acetate, ethanol, methanol, methyl acetate, 1,2-dichloroethane
unknown set consisting of three vials labelled #A, #B, #C
commercial fingernail polish removers

Safety

Safety glasses must be worn at all times in the laboratory. All organic liquids used in this experiment are to be considered harmful. Avoid ingestion of liquid or inhalation of vapors.

MOST ORGANIC LIQUIDS ARE HIGHLY FLAMMABLE. KEEP AWAY FROM HEAT SOURCES AND OPEN FLAME.

Experimental

Record all data in your notebook or on the **Results** sheet, whichever is applicable. Instructions for operation of the gas chromatograph and for performing syringe injections will be given during the prelab briefing. You will be provided with three vials, labeled with a number and the letters, A, B, and C. Each vial contains a pure substance or a mixture of pure substances from the following list:

Substance	Formula	Density (g/mL)	MW (g/mole)
acetone	C_3H_6O	0.7899	58.08
ethanol	C_2H_6O	0.7893	46.07
methanol	CH_4O	0.7914	32.04
ethyl acetate	$C_4H_8O_2$	0.9003	88.12
methyl acetate	$C_3H_6O_2$	0.9330	74.08
1,2-dichloroethane	$C_2H_4Cl_2$	1.2351	98.96

Inject a specified volume (see your instructor) of the contents of each vial into the GC and obtain the chromatogram. Identify the number of components and record retention time and signal area for each component. Through group discussion, devise a procedure for assigning the observed signals to pure substances from the above list. Find the number of moles of each component in the injection volume, based on signal areas.

Next, inject one of the commercial fingernail polish removers provided. Identify the substances present, and the active ingredient. Can you determine the approximate mole percent of active ingredient (what information is needed to determine this)?

Disposal Methods

Make sure all vials are tightly capped. Return them to the appropriate locations.

Questions

1. What experimental factors affect retention time?
2. What experimental factors affect signal intensity?
3. What other commercial products might you expect to be able to analyze using gas chromatography?

Preparation Mixtures

Name ________________
Section ________________
Date ________________

Read

1. This experiment.
2. The appropriate sections of your textbook.

Problems

1. Explain what is meant by the retention time of a substance.

2. What is the significance of the area under a GC signal?

3. A 2-μL sample of acetone is injected to a gas chromatograph. How many moles were injected?

Results
Mixtures

Name ________________
Section ________________
Coworkers ________________
____
____
____
Date ________________

Data

GC Parameters

Column temperature _____
Injector temperature _____
Detector temperature _____
Carrier gas flow rate _____
Attenuation factor _____

Analysis of vial contents

Unknown number ___

Vial	Retention Times	Area	Substances Present	Mole %
A				
B				
C				

Explanation of procedure used to assign observed signals to specific substances:

Analysis of fingernail polish remover

Product name_____
Active ingredient _____
Mole % active ingredient_____

Answers to Questions

Supplementary Optional Activities

(developed by DS Burns)

Reproducibility Check

A. Inject an appropriate volume of ethanol three or four times during a single two-minute GC run.
B. Obtain signal areas for all injections. Are the signal areas the same? Should they be? Calculate the average signal area for an injection of ethanol.

Analysis of Percent Ethanol in Whiskey

A. Seagram's Seven 7 Crown Whiskey is labeled as 40% alcohol by volume. Determine the actual percent alcohol in Seagram's Seven. You will have to determine a convenient injection volume.

Analysis of a Bic Lighter

A. Insert a sampling syringe into the Bic lighter, turn upside down, and draw out an appropriate volume of lighter fluid. Quickly inject into the GC. Make two more injections during the run.
B. Adjust signals to full scale.
C. Print the chromatogram and tabulate signal areas and retention times.

Analysis of Denatured Alcohol Solvent

A. Inject an appropriate volume of solvent.
B. Adjust signals to full scale.

1. How many components are in the mixture?
2. What alcohol is present in the solvent?

Analysis of a Gasoline

A. With help from your instructor, adjust the GC system parameters for analysis of the gas mixture:
B. Inject an appropriate volume of Texaco Unleaded 87 Octane gasoline and initiate the programmed run.
C. From the shape of the chromatogram, what conclusions can you make about the composition of gasoline?
D. Inject a mixture of pentane, hexane, octane, and decane as in step B. What similarities do you see between the resulting chromatogram and the chromatogram of gasoline?
E. How could you adjust the GC parameters to achieve a better separation of the gasoline mixture?

Results
Mixtures, Supplementary Optional Activities

Name ______________
Section ______________
Coworkers ______________
Date ______________

Data

1. Average signal area _____

2. % alcohol in Seagram's Seven _____

3. Retention time of butane _____
   Average signal area of injection _____

4. Number of components in mixture _____
   Alcohol present _____

5. What does the chromatogram tell you about the composition of gasoline?

Infrared Spectroscopy of Polymers

1 lab period; work in groups. Complete the Preparation page before laboratory.

Goals

- To understand the structures of polymers
- To use infrared spectroscopy to discover the identities of common plastics
- To discover the types of chemical groups attached to a polymer backbone

Background

Polymers are compounds consisting of extremely long chains of atoms. As the name implies, they are made up of many identical smaller units called monomers. Monomers are molecules of low molecular weight and usually contain a double bond. In this experiment, we will consider polymers built up from monomers of the general structure in Figure 1, where R is one of several simple chemical groups. For example, R can be hydrogen, H, a methyl group, H_3C, or a phenyl group, C_6H_5. The crucial structural feature of the monomer is the double bond between carbons, which consists of one strong σ and one relatively weak π bond. In the presence of other monomer molecules, the π bond breaks, and the carbon atoms form strong single bonds with similar carbons on neighboring monomer molecules. Figure 2 illustrates the linking of three monomers. Formation of a long chain results from linking many monomers in this fashion. The resulting polymer molecule has a periodic (i.e., repeating) structure in which R groups protrude from every other carbon of the hydrocarbon (i.e., carbon and hydrogen) backbone. Figures 3a and 3b represent different ways of drawing polymers. Figure 3a clearly shows the monomer origins of the polymer, while 3b, called a line structure, represents the hydrocarbon backbone as a zigzag line with attached R groups. Each line segment in Figure 3b represents a bond between carbon atoms. It is understood in this structure that each "kink" in the zigzag line represents a CH_2 group. Figure 4 illustrates another representation of the monomer. Double and single bonds and the R groups are shown explicitly. A CH_2 group at the left end of the double bond and a CH group at the right end are implied. The number of implied hydrogen atoms is determined by counting bonds to carbon atoms. A carbon atom (almost) always forms 4 bonds. Line structures emphasize the bonds present in a molecule.

Many commercial plastics are polymers of the type shown in Figure 3, with a simple hydrocarbon backbone and R groups attached to every or every other carbon. Since not only the hydrocarbon backbone but also the various R groups undergo characteristic vibrational motions, each different polymer, or plastic, will show a unique infrared spectrum. In this experiment, we will use infrared spectroscopy to address the following questions:

What polymers are present in commercial plastic films?
What types of R groups are present in these polymers?

Equipment and Materials

infrared spectrometer (preferably FT)
infrared spectra of commercial plastics
unknown polymer samples

Figure 1:

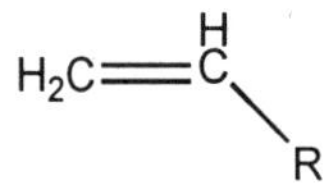

Figure 2:

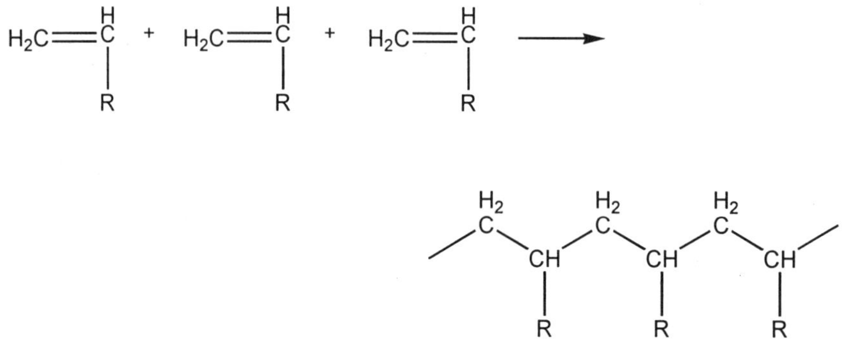

Figure 3:

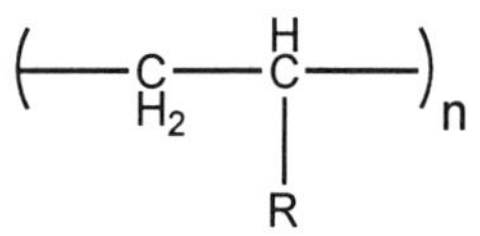

a

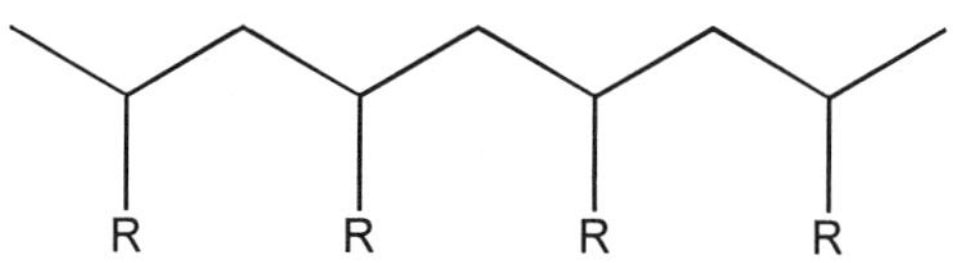

b

Figure 4

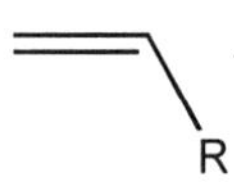

Safety

Safety glasses must be worn at all times in the laboratory.

Experimental

Record all data in your notebook or on the **Results** sheet, whichever is applicable.

Each group will identify several unknown plastic film samples. After receiving instruction in the use of the infrared spectrometer, obtain printed infrared spectra of your polymer samples, and compare the spectra with those of known plastics to make your identifications. When you have completed this, study the spectra of your samples. What can you learn about the kinds of chemical bonds present in the polymer structure? Look for absorptions in unusual regions of the IR spectrum: > 3000 cm^{-1}, between 1500 and 2800 cm^{-1}. Bands in these regions are distinctive because they stand out. The information in Table 1 might help you in identifying the types of bonds in the -R group attached to the polymer. Figure 5 shows -R groups that occur commonly in polymers.

Table 1
Characteristic Vibrational Frequencies

Region	Bond Type	Frequency, Intensity, Shape	Type of Vibration
4000-2700	OH	3600-2500 s, broad	stretch
	NH_2 or NH	3500-3300 w-m, broad	stretch
	=C(R)-H	3100-3000 w	stretch (C-H)
	R_3C-H	3000-2800 s	stretch (C-H)
2700-1850	C≡N	2300-2200 w	stretch
	C≡C	2100-2400 w	stretch
	C≡O	2200-2100 s	stretch
1850-1500	C=O	1780-1650 s	stretch
	C=N	1600-1500 s	stretch
	C=C	1670-1640 w-m	stretch
	NH_2	1650-1580 m-s, broad	bend

Key to abbreviations: s = strong, m = medium, w = weak. These refer to intensity of absorption.

Figure 5

Some R Groups in Polymers

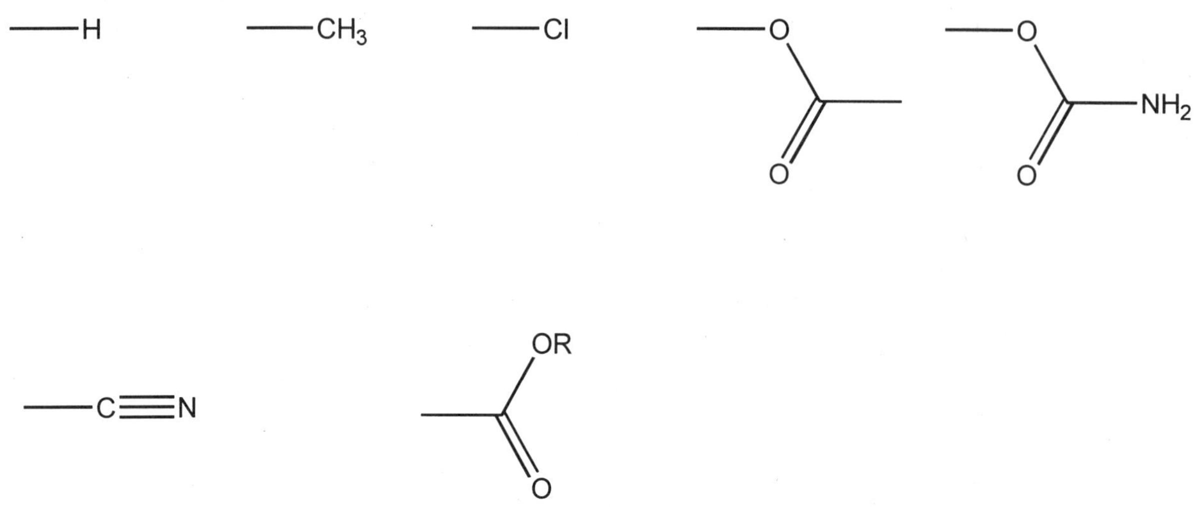

Disposal Methods

No disposal is required. Return the films to the instructor.

Questions

1. Indicate where each of the following polymers should have prominent bands in the IR spectrum:
    a. poly(acrylonitrile) R is CN
    b. poly(vinyl acetate) R is CH_2COOH
    c. poly(amide) R is $CH_2C(=O)NH_2$

Preparation
Infrared Spectroscopy of Polymers

Name ________________

Section ________________

Date ________________

Read

1. This experiment.
2. The appropriate sections of your text.

Problems

1. Draw the structure of each molecule using the simplified notation illustrated in Figures 3b and 4 above:

   a. Acetone, $CH_3C(=O)CH_3$

   b. butadiene, $CH_2=CH-CH=CH_2$

   c. isoprene, $CH_2=C(CH_3)-CH=CH_2$ (This is the monomer of natural rubber.)

   d. alanine, $HOOC-CH(CH_3)-NH_2$ (The amino acids are monomers of proteins.)

Results
Infrared Spectroscopy of Polymers

Name ________________
Section ________________
Coworkers ________________
____
____
____
Date ________________

Data

Unknown Polymer Samples

Sample Number	Identity	Frequencies of Prominent Bands (≥ 1500 cm^{-1})	Type of Bond	R

Answers to Questions

Stoichiometry

1 lab period; work in groups. Complete the Preparation page before laboratory.

Goals

- To learn the operation of an electronic absorption spectrometer
- To learn how to handle and care for spectrometer cells
- To learn methods for accurate dispensing of solutions
- To understand molarity
- To discover important relationships between masses of reactants and products in chemical reactions

Background

In this experiment, we explore some very important aspects of a key chemical concept, reaction stoichiometry. Consider the four reactions shown below:

Reaction 1: $xC_{12}H_8N_2 \cdot H_2O + FeSO_4 \cdot 7H_2O \rightarrow Fe(C_{12}H_8N_2)_x(SO_4) + 8H_2O$

$C_{12}H_8N_2$ = (o-phenanthroline)

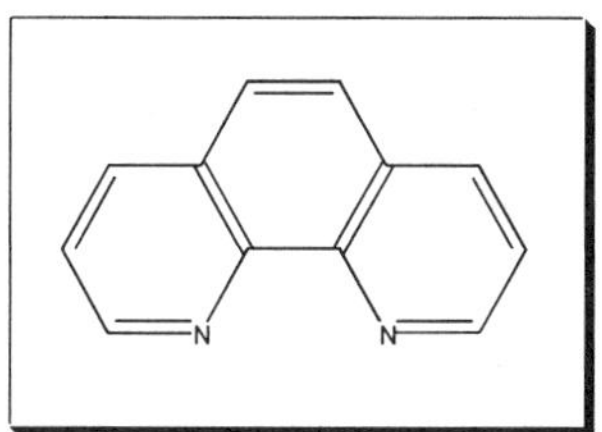

Reaction 2: $xH_4EDTA + CuCl_2 \cdot 2H_2O \rightarrow H_{4x-2}Cu(EDTA)_x + 2HCl + 2H_2O$

H_4EDTA = (ethylenediamine-tetraacetic acid)

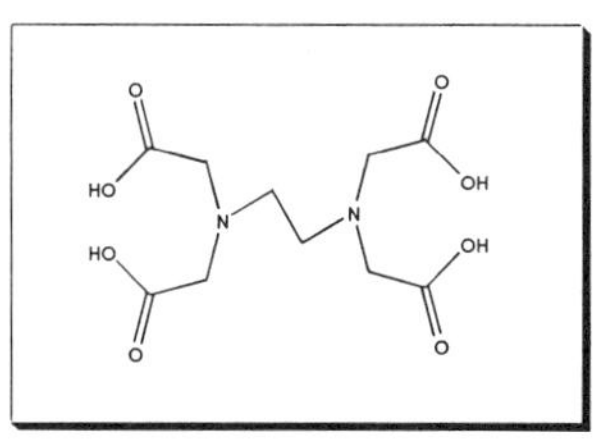

Reaction 3: $xH_4EDTA + CuSO_4 \cdot 5H_2O \rightarrow H_{4x-2}Cu(EDTA)_x + H_2SO_4 + 5H_2O$

Reaction 4: $xC_6H_5NO_3 + Ba(OH)_2 \cdot 8H_2O \rightarrow Ba(C_6H_4NO_3)_x + 10H_2O$
(p-nitrophenol)

We assign the designations in Table 1 to the reactants and products in the four reaction systems above. Molecular weights are given in the table.

Table 1

Reaction	Reagent 1	Reagent 2	Colored Product	Monitor at
1	$C_{12}H_8N_2 \cdot H_2O$	$FeSO_4 \cdot 7H_2O$	$Fe(C_{12}H_8N_2)_x(SO_4)$	510 nm
	MW=198.23	278.03		
2	H_4EDTA	$CuCl_2 \cdot 2H_2O$	$H_{4x-2}Cu(EDTA)_x$	320
	292.24	170.486		
3	H_4EDTA	$CuSO_4 \cdot 5H_2O$	$H_{4x-2}Cu(EDTA)_x$	320
	292.24	249.68		
4	$C_6H_5NO_3$	$Ba(OH)_2 \cdot 8H_2O$	$Ba(C_6H_4NO_3)_x$	460
	139.11	315.46		

For each reaction, we have designated the reactants as 1 and 2, and identified the colored product. The color is due to absorbtion of light in the visible region of the electromagnetic spectrum by the compound. It is possible to measure the concentration of colored product formed in solution by determining the amount of light that the solution absorbs. This is done using an instrument called an electronic absorption spectrometer. A spectrometer is an instrument used to measure the amount of electromagnetic radiation absorbed by a solution. The wavelength of interest determines the type of spectrometer used. For example, in this experiment, the electromagnetic radiation of interest is in the visible region so a simple Spectronic 20 and regular glass sample holders can be used.

The spectrometer consists of several key components. The light source must be stable over time because it must supply a constant amount of light to the sample and detector(s). A monochromator is used to select only the wavelength of interest and blocks out all other wavelengths. The detector measures the amount of light passing through the sample. First, the detector response is zeroed by measuring the signal generated by a blank. Then by comparing the signal with the sample present to the signal with the blank, the amount of light absorbed by the sample is converted to a signal displayed on an output.

Your section will investigate one of the reactions, 1-4. You will mix together specified known masses of the reactants (materials on the left side of the arrow) and monitor the formation of colored product using electronic absorption spectroscopy in order to answer the following question:

What is the value of x in your assigned reaction?

Equipment and Materials

electronic absorption spectrometer (or Spectronic 20)
1-cm sample cuvettes
8 Pasteur pipets
7 10-mL volumetric flasks
2 5-mL and 1 2-mL Mohr pipets

3 and 10 mL syringe pipet pumps or pipet bulbs
$5x10^{-4}$ M $C_{12}H_8N_2$
$3.33x10^{-4}$ M $FeSO_4 \cdot 7H_2O$
0.005 M H_4EDTA
0.01 M $CuCl_2 \cdot 2H_2O$
0.01 M $CuSO_4 \cdot 5H_2O$
0.0025 M p-nitrophenol
0.0025 M $Ba(OH)_2$

Safety

Safety glasses must be worn at all times in the laboratory. All reagents used in this experiment should be considered harmful. Avoid contact with the skin. Clean up all spills immediately, using plenty of water.

Experimental

Record all data either in your lab notebook or on the **Results** sheet, whichever is applicable. Instructions for handling cells and for operation of the electronic absorption spectrometer will be given in class.

Your instructor will indicate which of the reaction systems your section is to study Each group will be given an equipment box and assigned seven different volumes of reagent 2 to react with the specified volume of reagent 1. In a clean, dry beaker (50- or 100-mL), obtain about 40 mL of reagent 1 solution. In a second clean, dry beaker (50- or 100-mL), obtain a volume of reagent 2 solution equal to the sum of your assigned volumes + 5 mL. Pipet 5 mL of reagent 1 solution and the first assigned volume of reagent 2 solution into a 10-mL volumetric flask and add distilled water to the mark on the flask. Mix well. Obtain the electronic absorption spectrum of the solution and record the absorbance at the wavelength specified in Table 1. Repeat for the six other specified volumes of reagent 2. The instructor will discuss how your data are to be handled.

Disposal Methods

Pour all solutions containing Cu or Ba in the waste bottle labelled "metal waste." Solutions of p-nitrophenol and o-phenanthroline should be poured in the waste bottle labelled "organic waste." All other solutions should be poured in the sink and rinsed down the drain with plenty of water.

Questions

While waiting for the postlab discussion, work on the following questions.

1. (To be answered after you have received your section data). What is the <u>uncertainty</u> associated with any particular absorbance value in "region 2" of the absorbance-volume plot? Based on class data, what is the <u>precision</u> in the limiting absorbance value in region 2?
2. Write a paragraph explaining the following aspects of the section plot of absorbance versus volume of reagent 2:
    a. The linear region with positive slope.
    b. The linear region with zero slope.
    c. The point of intersection of the two linear regions.

Preparation
Stoichiometry

Name ____________________
Section ____________________
Date ____________________

Read

1. **Uncertainty, Precision, and Error in Experimental Data**, in this manual.
2. The appropriate sections of your text.

Problems

1. Consider a plot of absorbance versus the volume of reagent 2 solution added to a particular volume of reagent 1 solution for any of the reactions 1-4. What would such a plot look like? Sketch it.

2. For reaction 1, what volume of reagent 2 should be required to react with 5.0 mL of reagent 1, assuming that x is 1? 2? 3?

Results
Stoichiometry

Name ____________________
Section ____________________
Coworkers ____________________
____
____
____
Date ____________________

Data

Reaction:

Wavelength: ____nm

Reagent 1: Concentration of stock solution ________

Volume ______
Moles ______
Mass ______

Reagent 2: Concentration of stock solution ________

Volume 1 ______
Moles ______
Mass ______
Absorbance ______

Volume 2 ______
Moles ______
Mass ______
Absorbance ______

Volume 3 ______
Moles ______
Mass ______
Absorbance ______

Volume 4 ______
Moles ______
Mass ______
Absorbance ______

Volume 5 ______
Moles ______
Mass ______
Absorbance ______

Volume 6 ______
Moles ______
Mass ______
Absorbance ______

Volume 7 ______
Moles ______
Mass ______
Absorbance ______

Value of x in reaction ____

Answers to Questions

Vapor Pressure

1 lab period; work in groups. Complete the Preparation page before laboratory.

Goals

- To learn techniques of headspace gas chromatography
- To discover the manner in which the vapor pressure of a pure liquid varies with temperature

Background

The pressure of vapor in equilibrium with a liquid in a closed container--the equilibrium vapor pressure of the liquid--is a simple example of an equilibrium constant. Consider a liquid in a closed container; e.g., acetone in a small, stoppered Erlenmeyer flask. Acetone evaporates to some extent within the container. Therefore acetone vapor is present over the liquid, as shown in Figure 1.

After some time, the number of molecules in the vapor phase is sufficient to give equal rates of evaporation, in which molecules escape the surface of the liquid, and condensation, in which molecules from the vapor re-enter the liquid. The amount of vapor and therefore its pressure then remains constant with time. The equalization of R_E (rate of evaporation) and R_C (rate of condensation) is shown graphically in Figure 2. When the rates are equal, the pressure of vapor no longer changes with time, and we say that equilibrium exists between liquid and vapor. The pressure due to vapor at equilibrium is called the equilibrium vapor pressure of the liquid, P_{vap}.

Figure 1

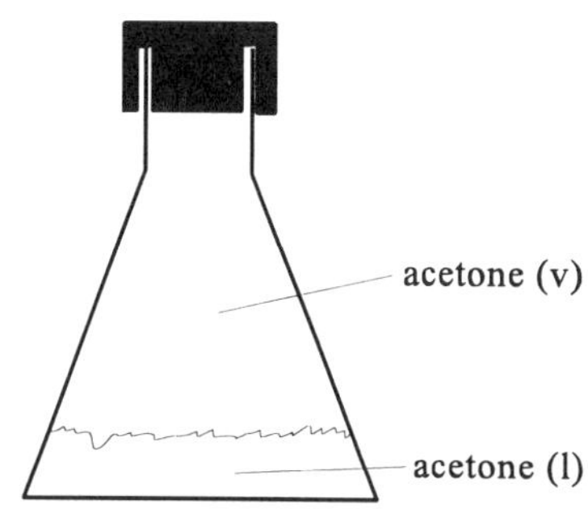

We would expect P_{vap} to depend on the temperature through the following line of reasoning:

- To leave the liquid phase, molecules must overcome forces of attraction from other liquid molecules (i.e., they must climb out of a potential well);
- Whether they can do this depends on how much kinetic energy (KE) they have;
- The average KE per molecule increases with T (KE(molecule) = 3/2 kT);
- Therefore the number of molecules able to escape the liquid increases with T, and P_{vap} increases with T.

In this experiment we will explore the dependence of P_{vap} on temperature. We will use headspace gas chromatography to measure the vapor pressures of three pure liquids at several temperatures. Our goal will be to discover answers to the following questions:

Figure 2

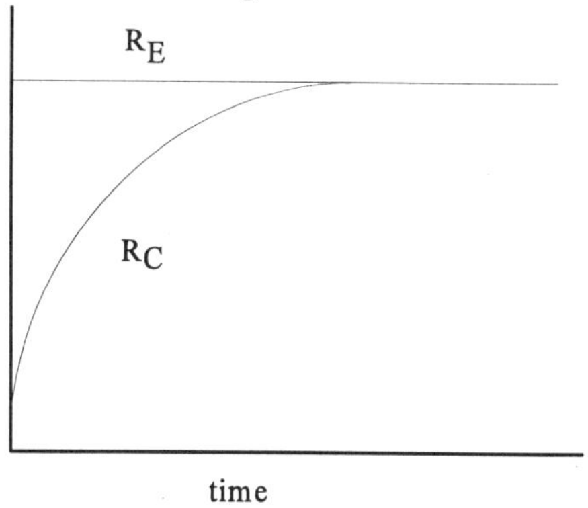

Is a plot of P_{vap} versus T linear, curved with an increasing slope, or curved with a decreasing slope? If the P_{vap}-T plot is not linear, is there a way to linearize it?

Equipment and Materials

gas chromatographs
2.0-mL gas tight syringes
25-mL Erlenmeyer flasks with septum caps
constant temperature bath
reagents: acetone, ethanol, hexane

Safety

Safety glasses must be worn at all times in the laboratory. All liquids used in this experiment are to be considered harmful. Avoid ingestion of liquid and inhalation of vapor.

MOST ORGANIC LIQUIDS ARE HIGHLY FLAMMABLE. KEEP AWAY FROM HEAT SOURCES AND OPEN FLAME.

Experimental

Record data in your lab notebook or on the **Results** sheet, whichever is applicable. One of the three liquids and a gas chromatograph (GC) will be assigned to your group. Liquid samples are contained in septum-capped 25-mL Erlenmeyer flasks clamped in a constant temperature bath. After receiving instructions in headspace sampling technique and in GC use, send one member of your group to the bath. Record the bath temperature. Sample 0.2 mL of headspace vapor using the gas tight syringe and dilute the vapor sample with 0.6 mL of air (these volumes are subject to modification by your instructor). Inject the sample slowly and smoothly into the GC port while starting data collection. Make at least 6 additional injections of headspace vapor during an ~ 8-minute GC run so that you obtain several (at least three) signals whose areas reproduce to within ±3%. Print your chromatogram. When everyone has finished runs at a given bath temperature, the instructor will set the bath for a second temperature. Repeat the GC injection procedure at each new bath temperature until you have data at four temperatures in the range -10 to 20°C.

Determine the mean signal area at each temperature, and divide this area by the mean area at the highest temperature (20°C) to obtain relative vapor pressures. (These calculations may require correction for the GC attenuation factor you recorded for each temperature. Attenuation factors are related by powers of 2 to make this correction simple: a signal whose area is 40 when attenuation is 16 will have an area of 20 at attenuation 32.) The instructor will indicate how your data are to be handled. Data will be analyzed in the postlab discussion.

NOTE: There will be times during which some groups are not working at the GCs. In addition, bath temperature changes require time. Use the free time to record data on the **Results** sheet, do the average area calculations, and work on **Questions** 1-4.

Disposal Methods

No disposal is required.

Questions

1. Consider the system in Figure 3, consisting of a liquid in equilibrium with its vapor in a syringe at some temperature. The plunger of the syringe is stationary. On graphs like those in the figure, show how the <u>rate of evaporation</u> and <u>rate of condensation</u> change with time as a result of each process described below.

a. From t_o to t_1, the system is at equilibrium.
b. At t_1, the plunger is instantly moved from point A to point B. The system comes to a new equilibrium.
c. At t_2, the temperature of the system is instantly increased. The system comes to equilibrium at the new temperature.

Figure 3

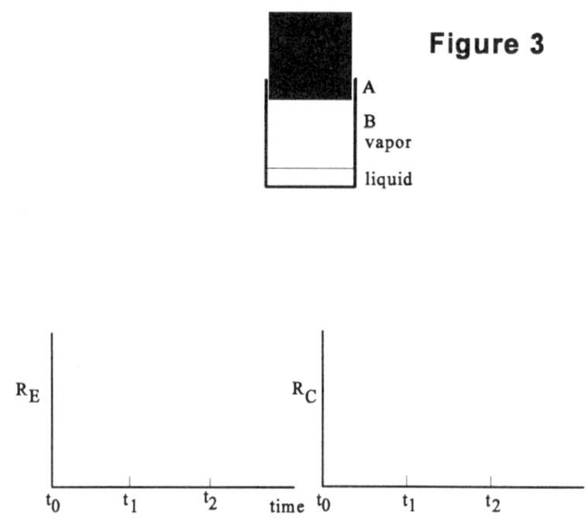

2. (Answer based on section data.) Is your P_{vap} vs. T(K) plot consistent with the ideal gas law? Explain.
3. (Answer based on section plot.) Can you offer an explanation in terms of molecular behavior for your P_{vap}-T(K) plot?

Preparation
Vapor Pressure

Name __________________
Section __________________
Date __________________

Read

1. This experiment
2. The appropriate sections of your textbook

Problems

1. In analyzing liquids by GC, it is necessary to inject only about 1 μL of sample. Why do we have to inject such a large volume of sample to the GC in this experiment?
2. A 0.2-mL injection of heptane vapor at -10 °C produces a signal with area 7.9 at attenuation 16. A similar injection of heptane vapor at 20 °C produces a signal with area 29.7 at attenuation 32. What is the vapor pressure of heptane at -10 °C relative to that at 20 °C (corrected for attenuation)?
3. Consider the system shown in Figure 4, consisting of 1 mL acetone liquid in a septum-capped 25-mL Erlenmeyer flask. The liquid acetone is in equilibrium with acetone vapor at pressure P_{vap} at 25 °C. What effect will each of the following have on the value of the equilibrium P_{vap}?

   a. T is instantly raised to 40°C.
   b. Some air is injected into the flask with a syringe.
   c. Some acetone vapor is withdrawn from the flask with a syringe.
   d. T is instantly lowered to 15°C.
   e. The liquid is withdrawn by syringe and injected through a septum cap into a 50-mL flask.
   f. Another 1 mL of acetone liquid is injected into the flask.

Figure 4

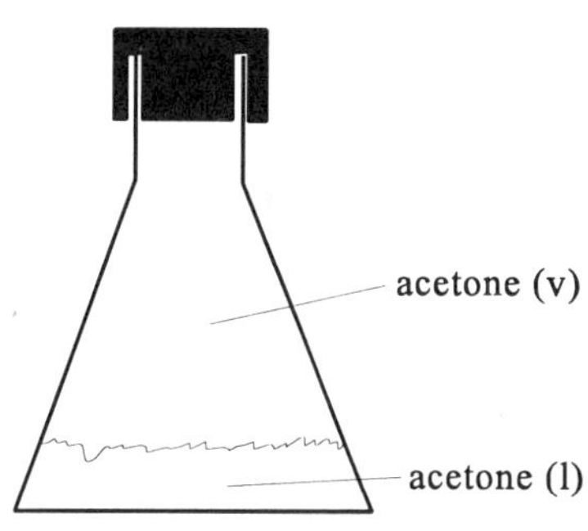

Results
Vapor Pressure

Name ______________
Section ______________
Coworkers ______________
____
____
____
Date ______________

Data

Liquid:________
Vapor Pressure Data

T(K)	Injection #	Signal Area	Relative P_{vap}	Attenuation
_____	_____	_____	_____	_____
	_____	_____	_____	_____
	_____	_____	_____	_____
	_____	_____	_____	_____
	_____	_____	_____	_____
	Mean:	_____	_____	
_____	_____	_____	_____	_____
	_____	_____	_____	_____
	_____	_____	_____	_____
	_____	_____	_____	_____
	_____	_____	_____	_____
	Mean:	_____	_____	
_____	_____	_____	_____	_____
	_____	_____	_____	_____
	_____	_____	_____	_____
	_____	_____	_____	_____
	_____	_____	_____	_____
	Mean:	_____	_____	
_____	_____	_____	_____	_____
	_____	_____	_____	_____
	_____	_____	_____	_____
	_____	_____	_____	_____
	_____	_____	_____	_____
	Mean:	_____	_____	

Mean: ______

ΔH°_{vap} = ______

Answers to Questions

Interaction of Light and Solutions

1 lab period; work in pairs. Complete the Preparation page before laboratory.

Goals

- To find a relationship between absorbance and concentration
- To find the concentration of an unknown solution

Background

Electromagnetic radiation from x-rays into the microwave region can be absorbed by molecules. The wavelength, thus the energy, absorbed by a molecule depends on the structure of the molecule. When the absorbed radiation is between about 400 and 700 nm, we perceive that the absorbing solution has color. The color absorbed, however, is the complement of the observed color. For example, if the absorbance is around 410 nm, this is "violet" light but a solution absorbing these wavelengths will appear yellow. When all of the visible light is absorbed, an object appears black while if light between 400 and 700 nm is not absorbed, a solution appears colorless. However, many colorless solutions absorb radiation in the ultraviolet range, wavelengths below 400 nm, and/or in the infrared region beyond 700 nm.

A spectrometer is an instrument used to measure the amount of electromagnetic radiation absorbed by a solution. The wavelength of interest determines the type of spectrometer used. For example, in this experiment, the electromagnetic radiation of interest is in the visible region so a simple Spectronic 20 and regular glass sample holders, called cuvettes, can be used. However, special instrumentation and ultrahigh vacuum conditions are required to study the absorbance of x-rays. To study absorbance of ultraviolet light, special quartz glass cuvettes and an arc light source are needed. Since glass interferes with measurements in the infrared region, sample cells with salt plate windows are used.

A spectrometer consists of several key components. The light source must be stable over time supplying a constant amount of light to the sample and detector(s). A monochromator selects only the wavelength of interest and blocks out all other wavelengths. Detectors measure the amount of light passing through the sample. First, the detector response is zeroed by measuring the signal generated by a blank. Then the signal generated with the sample present is compared to the blank. The amount of light absorbed by the sample is converted to an output to a recorder or meter. In some spectrometers, the detector is an array of small detectors each sensitive to a particular wavelength of light. The entire spectrum is scanned at once. A computer takes the signals from the detector array and displays the entire spectrum as absorbance vs. wavelength.

An important consideration when measuring solution absorbance is correcting for the background absorbance. In this experiment, the dye is dissolved in distilled water so an appropriate blank is simply distilled water. However, for more complex solvent systems, the blank would be prepared from all of the components of the solvent without the dye. For example, if the dye was dissolved in 3 M HCl and 2 M NaCl, the blank would be a solution of 3 M HCl and 2 M NaCl. In this way, you correct for any decrease in the light reaching the detectors caused by the solvent.

You will start with a dye solution of known concentration. Your group will be assigned a wavelength to study between 490 and 560 nm. By quantitatively diluting the stock solution, you can prepare a wide range of solutions of known concentrations. Measure the absorbance of each solution at the assigned wavelength. At least four solutions should have absorbance readings between 0.1 and 0.8. This may require some trial and error. Deduce a relationship between the absorbance and the concentration. Once you have found a relationship between absorbance and concentration, determine the concentration of dye in your unknown. Take one of your standard solutions and obtain an absorbtion spectrum for the sample and determine λ_{max}.

What is the relationship between absorbance and concentration?
How does one find the concentration of an unknown solution?
What factors determine the maximum and minimum absorbance you can accurately measure?

Equipment and Materials

spectrometer (Spectonic 20 or other)
volumetric pipets (1, 2, 5, 10-mL)
volumetric flasks (10, 25, 50, 100-mL)
stock erythrosin B (0.5 g/L)

Safety

Eye protection must be worn in the laboratory. The dye will stain skin and clothes.

Experimental

Record all data in your lab notebook or on the **Results** sheet, whichever is applicable. Prepare a cuvette with the blank. Be sure to place it into the spectrometer with the same orientation each time you zero the spectrometer. Prepare dilutions of the stock solution. Measure the absorbance of each standard at the assigned wavelength. The concentrations that can be measured will vary with the wavelength so this part of the experiment requires some trial and error. This is not unlike what you would encounter in a real life situation where the appropriate concentration range would be found experimentally after determining the optimum wavelength. Once you have obtained data, find a relationship between the absorbance and the concentration. Try to use as much of the available absorbance range of the spectrometer as possible. Once you have a relationship, apply it to the unknown. Measure the absorbance of the unknown and find the concentration based on your relationship.

Disposal Methods

Place all used solutions in the waste beakers or bottles in the lab. Do not flush solutions down the sink.

Questions

1. Draw a graph showing the relationship you found between absorbance and concentration.
2. How does one find the concentration of an unknown solution?
3. What is the concentration of the unknown solution?
4. From the class data, at what wavelength would you find the most sensitivity?
5. Why is it not necessary to determine a spectrum for each standard solution used to prepare the calibration curve?

Preparation
Interaction of Light and Solutions

Name ____________
Section____________
Date ____________

Read

1. This experiment.
2. About electromagnetic radiation in your text.

Questions

Define the following terms:

electromagnetic radiation
color wheel
absorbance
molar absorptivity

Practice Problems

From the stock dye (0.5 g/L) describe how you would quantitatively prepare solutions of the following concentrations and volumes:

concentration	total volume
0.1 g/L	100.00 mL
0.002 g/L	25.00 mL
0.05 g/L	50.00 mL

Results
Interaction of Light and Solutions

Name ______________
Coworker(s) ______________
____
Section ______________
Date ______________

Data

Wavelength ________

Table 1. Dilution, concentration and absorbance of dye solutions.

volume of stock	total volume	new concentration	absorbance

Concentration of unknown _______

Answers to Questions

Equilibrium

1 period; work in groups. Complete the Preparation page before laboratory.

Goals

- To measure an equilibrium constant using electronic absorption spectroscopy
- To discover how the equilibrium constant varies with temperature

Background

Salts show a large range of solubilities in water. Some, such as NaCl, dissolve to a large extent. Others, such as AgCl, dissolve only slightly. AgCl is an example of a sparingly soluble salt. When put in contact with water, such salts dissolve to produce small concentrations of their ions. When the maximum amount possible of the salt has dissolved, an equilibrium exists between the solid and dissolved forms of the salt, as shown for AgCl in equation 1.

$$AgCl(s) \rightarrow Ag^+(aq) + Cl^-(aq) \qquad (1)$$

The equilibrium constant expression for this reaction is in equation 2.

$$K_{sp} = [Ag^+][Cl^-] = 1.8 \times 10^{-10} \qquad (2)$$

Since AgCl is present as a pure solid, its concentration is constant and is not written explicitly. The subscript label "sp" stands for solubility product. This conveys that the equilibrium constant is a product of ion concentrations closely related to the solubility of the salt in water.

There are many salts in which the cation/anion stoichiometry is not 1:1. These salts dissolve to give ion concentrations reflecting the stoichiometry. Equation 3 shows an example, and equation 4 shows the equilibrium constant expression.

$$Fe_2S_3(s) \rightarrow 2Fe^{3+}(aq) + 3S^{2-}(aq) \qquad (3)$$

$$K_{sp} = [Fe^{3+}]^2[S^{2-}]^3 \qquad (4)$$

Silver acetate, $AgC_2H_3O_2$, dissolves in water according to equation 5. Although neither silver ion nor acetate ion is colored, Ag^+ reacts quantitatively with a mixture of m- and p-phenylenediamine according

$$AgC_2H_3O_2(s) \rightarrow Ag^+(aq) + C_2H_3O_2^-(aq) \qquad (5)$$

to equation 6 to produce a product that absorbs light in the visible region of the spectrum. By measuring the amount of light absorbed by this product, you can calculate the amount of Ag^+ present in a saturated solution of $AgC_2H_3O_2$. The stoichiometry of equation 5 requires that $[Ag^+] = [C_2H_3O_2^-]$. Your goal in this experiment is to answer the following questions:

How can K_{sp} for silver acetate be measured?
How does K_{sp} for silver acetate depend on temperature?

m-phenylenediamine =

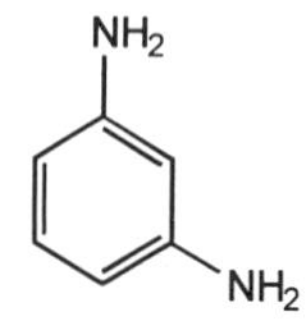

p-phenylenediamine =

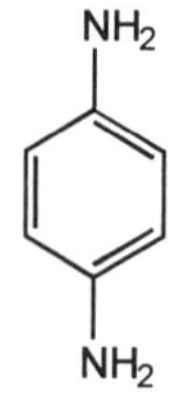

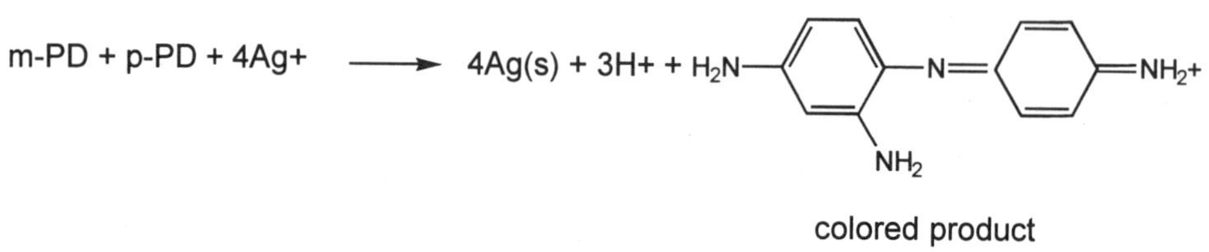

(6)

Equipment and Materials

electronic absorption spectrometer (or Spectronic 20)
10-mL volumetric flasks with caps
50-mL volumetric flask with stopper
spectrometer cells
5-mL measuring pipet
2-mL measuring pipet
5 1-mL measuring pipets
Pasteur pipets
4 5- or 10-mL graduated cylinders
saturated silver acetate solution
pH 7 buffer solution
2 x 10^{-4} M $AgNO_3$ solution
m-phenylenediamine solution (1 g/L)
p-phenylenediamine solution (1 g/L)

Safety

Safety glasses must be worn at all times in the laboratory. All solutions are considered harmful. Avoid ingestion and contact of these solutions with the skin. Wash all spills immediately with plenty of water. Silver salts stain the skin black. Such discoloration wears off after a couple of days.

Experimental

Record all data in your lab notebook or on the **Results** sheet, whichever is applicable. Instructions for handling cells and using the electronic absorption spectrometer will be given in class. There are 2 parts to the experiment. In the first part, the class will construct a standard curve by measuring the amount of colored product formed by known concentrations of Ag^+. In the second part, the concentration of colored product will be used to determine the amount of Ag^+ in a saturated solution of silver acetate.

Part 1. Preparation of Standard Curve

Your team will be assigned 2 volumes of standard $AgNO_3$ solution to investigate. In 5- or 10-mL graduated cylinders, obtain 1 mL more than the total required volume of standard $AgNO_3$ solution; 5 mL pH 7 phosphate buffer; 5 mL m-phenylenediamine solution; and 5 mL p-phenylenediamine solution. To a 10-mL volumetric flask, pipet the first assigned volume of $AgNO_3$ solution, 1.00 mL of pH 7 buffer, and 1.00 mL m-phenylenediamine solution. Then pipet 1.00 mL p-phenylenediamine solution into the flask, noting the exact time the last drop enters the flask. Add distilled water to the mark and mix well. BE SURE TO ADD THE REAGENTS TO THE FLASK IN THE SPECIFIED ORDER. Allow to stand for eight minutes, then transfer some of the solution to a spectrometer cell. Exactly 10 minutes after you finished adding the p-phenylenediamine, run the electronic absorption spectrum and obtain the absorbance at 550 nm. Repeat for the second assigned volume of $AgNO_3$ solution. Your instructor will indicate how you are to handle your data. Work on questions while waiting for midlab discussion.

Part 2. Determination of K_{sp} for Silver Acetate

Located in the lab are constant temperature baths containing saturated solutions of silver acetate thermostatted in the range 0-50 °C. Your team will be assigned one of these temperatures to study. Find the appropriate bath. Put about 25 mL distilled water into a 50-mL volumetric flask. Pipet 0.50 mL of saturated silver acetate solution into the flask and add distilled water to the mark. Stopper and mix. To a 10-mL volumetric flask, IN THE ORDER SPECIFIED, pipet 1.00 mL of this solution; 1.00 mL of pH 7 buffer; and 1.00 mL m-phenylenediamine. Finally, add 1.00 mL p-phenylenediamine, noting the exact time at which the last drop enters the flask. Dilute to the mark with distilled water. After 8 minutes, transfer some of the solution to a spectrometer cell, then run the spectrum exactly 10 minutes after you started timing. Measure the absorbance at 550 nm. Repeat with a second aliquot from the 50-mL volumetric flask. Calculate the average K_{sp} for silver acetate. Your instructor will indicate how the data are to be handled.

Disposal Methods

Pour all solutions in the designated waste bottles.

Questions

1. AgCl is placed in contact with 100.0 mL of distilled water. What mass of AgCl will dissolve in the water?

References

1. "A Semimicro Spectrophotometric Determination of the Ksp of Silver Acetate at Various Temperatures" by John Liebermann and Ki J. Yun. *Journal of Chemical Education*, **1988**, 65, 728-729.

Preparation
Equilibrium

Name ________________
Section ________________
Date ________________

Read

1. This experiment.
2. The appropriate sections of your textbook.

Problems

1. Suppose you place 1.0 g $AgC_2H_3O_2$ in contact with 50 mL of distilled water and gently stir the mixture. Show graphically how the concentrations of Ag^+ and $C_2H_3O_2^-$ in solution will change with time, assuming that solid and liquid are placed in contact at t = 0. The plot should be qualitative.

2. Describe what happens at the molecular level once equilibrium is reached--that is, once the solution is saturated with $AgC_2H_3O_2$.

Results
Equilibrium

Name ______________

Coworkers ______________

____

____

____

____

Section ______________

Date ______________

Data

Standard Curve

Volume Standard $AgNO_3$	$[Ag^+]$, moles/L	Absorbance at 550 nm

---

K_{sp} Determination

Temperature	Absorbance at 550 nm

Calculations:

$[Ag^+]$ in Saturated $AgC_2H_3O_2$ Solution =____

K_{sp} = _____

Answers to Questions

Indicators

1 period; work in groups. Complete the Preparation page before laboratory.

Goals

- To gain experience in the use of the electronic absorption spectrometer
- To understand indicator color changes and their relationship to the equilibrium constant for reaction of the indicator with water
- To determine the equilibrium constant, K_{In}, for a particular indicator
- To use the indicator to signal the equivalence point in an acid-base titration

Background

An acid-base indicator is a weak acid that reacts with water, like any weak acid, according to the following general reaction:

$$HIn + H_2O \rightarrow H_3O^+ + In^- \qquad K_{In}$$

The usefulness of an indicator lies in the fact that one (or both) of the forms, HIn or In^-, is intensely colored. The appearance, disappearance, or change of indicator color in a solution during a titration of an acid with a base may be used as a signal that a certain pH has been reached, and that it is time to stop the titration. An indicator is chosen for a titration so that its color change coincides with the equivalence point of the titration--the point at which the amount of added base (acid) is stoichiometrically equivalent to the amount of acid (base) originally present in the solution. The equivalence point in the titration of an acid with a base occurs at pH 7 or higher; when a base is titrated with an acid, the equivalence point has pH 7 or lower. Generally, an indicator changes color over a pH range of 2 units, centered about the value of pK_{In} (-log K_{In}). For example, the indicator thymolphthalein, HTP, has a pK_{In} of 10.0. Its acid form, HTP, is colorless, and its base form, TP^-, is blue. A change from colorless to blue occurs over the pH range 9 to 11, and is useful in titrations of acids with base. The indicator alizarin red S, HARS, has a pK_{In} of 5.4. Its acid form is red and its base form, ARS^- is yellow. It undergoes a change from yellow to red over the pH range 6.4 to 4.4, and is useful in titrations of bases with acid. Clearly, thymolphthalein would not be useful as an indicator for a titration where the pH at the equivalence point is 6, because at this pH, it would not have begun to change color. Similarly, alizarin red S would not be useful as an indicator for a titration where the pH at the equivalence point is 8, because it would undergo a color change long before this pH was achieved. The point in a titration where the indicator color change occurs is called the endpoint of the titration. It is essential in choosing an indicator for a titration that the endpoint coincide with the equivalence point.

In this experiment, your section will be assigned one of the indicators, methyl red (HMR) or phenolphthalein (HPP). Use electronic absorption spectroscopy to discover the answers to the following questions:

What is pK_{In} for your assigned indicator?
Can the indicator be used in a significant titration?

Equipment and Materials

electronic absorption spectrometer
1-cm sample cells
10-mL volumetric flasks
1-mL measuring pipets
2-mL measuring pipets
syringe pumps
Pasteur pipets/bulb
50-mL burets
stock solution of phenolphthalein
stock solution of methyl red
glycine buffers covering the pH range ≈7-11
citrate buffers covering the pH range ≈2.75-6.75
Vitamin C tablets
standard NaOH, ≈0.1 M
household ammonia cleaner
standard HCl, ≈0.1 M

Safety

Safety glasses must be worn at all times in the laboratory. You will work with solutions of acids and bases. Avoid ingestion and contact of these solutions with the skin. If you spill an acid or base solution on your skin, flush the area immediately with plenty of water.

Part 1. Determination of the pK_{In} of an Acid-Base Indicator

Experimental

Record all data in your lab notebook or on the **Results** sheet, whichever is applicable. Your instructor will tell you which indicator to investigate, and will assign your group 4 pH values to investigate. Obtain an equipment box and 4 buffers. Obtain no more than 3 mL indicator solution in the 5-mL graduated cylinder in your equipment box. Pipet 0.50 mL of indicator stock solution into each of your four 10-mL volumetric flasks. Fill the first volumetric flask to the mark with one of the buffer solutions, and label the flask with the pH. Quickly obtain the electronic absorption spectrum of the solution in the wavelength range 700-300 nm. Read and record the absorbance at 520 nm for methyl red or 552 nm for phenolphthalein. Fill the second volumetric flask to the mark with the second buffer solution, label, and obtain the spectrum and required absorbance value. Repeat for the third and fourth buffer solutions. When you are finished collecting data, process it as indicated by the instructor. While waiting for the midlab discussion, work on the **Questions**. Put answers on the **Results** sheet or in your notebook.

Disposal Methods

Pour used solutions into the sink, and flush down the drain with plenty of water.

Questions

1. Write the equilibrium constant expression for an indicator, HIn, so as to show directly what factors influence the concentration ratio, $[In^-]/[HIn]$.
2. What is the value of this concentration ratio when $pH = pK_{In}-1$; when $pH = pK_{In}$; when $pH = pK_{In}+1$?
3. Write a paragraph explaining why an indicator undergoes a color change over the pH range $pK_{In} \pm 1$.
4. Suppose we examine the electronic absorption spectrum of an indicator over a pH range including pK_{In}. How can we use these spectra to determine the value of pK_{In}?

Part 2. Use of an Indicator in a Practical Acid/Base Analysis

Experimental, Titration of Vitamin C Tablets

Obtain 60 mL standard NaOH (≈0.1 M) and two vitamin C tablets. Rinse a 50-mL buret with two 5-mL portions of standard NaOH solution, then fill the buret with NaOH solution. Using a mortar and pestle, thoroughly crush and grind the vitamin C tablets. Measure out about 0.25 g of powder into each of three 250-mL Erlenmeyer flasks. The masses should be known to 4 decimal places. Add about 75 mL distilled water to the first sample. Stopper the other two flasks. Swirl the first flask for 3-5 minutes to achieve maximum dissolution of the powder (the binders and fillers used in the tablet will not dissolve). Add 3 drops of phenolphthalein to the flask and titrate with standard NaOH from the buret until the solution has a persistent faint pink color (the solid in the flask makes the pink color difficult to see).

Titrate the other two samples in the same manner.

Calculate the amount of ascorbic acid in each sample, the weight percent of ascorbic acid in the sample, and the average weight percent of ascorbic acid in a vitamin C tablet. Estimate your precision. The formula and structure of ascorbic acid are given below.

Ascorbic acid formula: $HC_6H_7O_6$

structure:

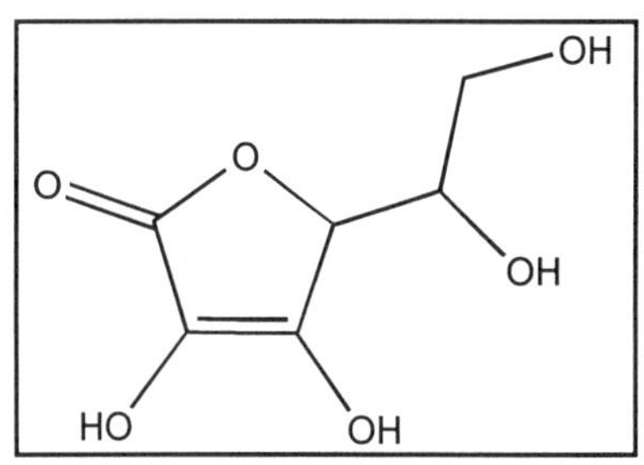

Experimental, Titration of Household Ammonia

Obtain 7 mL of household ammonia solution. Obtain 60 mL standard HCl solution (≈0.1 M). Rinse a 50-mL buret with two 5-mL portions of standard HCl solution then fill the buret with standard HCl solution. Pipet 2.00 mL of ammonia cleaner into each of three 250-mL Erlenmeyer flasks, and add about 25 mL distilled water. Stopper two flasks. To the other flask, add 5 drops methyl red solution, and titrate with standard HCl (≈ 0.1M) from the buret until the solution has a persistent pink color.

Titrate the other samples in the same manner.

Calculate the concentration of ammonia in each sample in moles per liter of cleaner, and the average concentration of the three samples. Estimate your precision.

Preparation
Indicators

Name ____________________
Section ____________________
Date ____________________

Read

1. This experiment.
2. The appropriate sections of your textbook.

Problems

1. An indicator, HIn, has $K_{In} = 4.2 \times 10^{-8}$. Over what pH range will it change color?

2. The following absorbance data are obtained as a function of pH for an unknown indicator. From the data, determine the value of K_{In} for the indicator.

pH	A
5.10	0.047
5.30	0.072
5.50	0.110
5.70	0.163
5.90	0.235
6.10	0.327
6.30	0.434
6.50	0.546
6.70	0.653
6.90	0.745
7.10	0.817
7.30	0.871
7.50	0.908
7.70	0.933
7.90	0.951
8.50	0.973

Results
Indicators

Name ________________
Section ________________
Coworkers ________________
____
____
____
Date ________________

Data, Part 1

Indicator__________ λ_{max} ____

Absorbance Data

Solution	pH	Absorbance at ____ nm
1		
2		
3		
4		

pK_{In} of indicator____

Data, Part 2

Analysis of Vitamin C Tablets

commercial brand of tablet_______

molarity of standard NaOH_______

	Trial 1	Trial 2	Trial 3
mass of sample			
final buret reading			
initial buret reading			
volume of NaOH used			
mass ascorbic acid			
weight % ascorbic acid			
mean weight % ascorbic acid in tablet			
standard deviation			

Analysis of Household Ammonia

commercial name of cleaner_________
molarity of standard HCl ____________

	Trial 1	Trial 2	Trial 3
volume of sample			
final buret reading			
initial buret reading			
volume of HCl used			
moles per liter of NH_3			
mean molarity of NH_3 in household cleaner			
standard deviation			

Answers to Questions

Intermolecular Forces

1 lab period; work in groups. Complete the Preparation page before laboratory.

Goals

- To learn the techniques of headspace gas chromatography
- To gain experience in accurate, quantitative dispensing of liquids
- To discover the manner in which the vapor pressure of a liquid varies with its environment

Background

Gas chromatography is a simple, fast, and accurate way to analyze mixtures of volatile liquids. It can be used similarly to analyze mixtures of gases, the difference being that larger sample volumes must be injected to get observable signals. In this experiment, we will use gas chromatography to measure the vapor pressures over pure liquids and binary mixtures of liquids in closed containers, to explore the relationship between the composition of the vapor and that of the liquid. Gas chromatography of a sample of vapor from the headspace over a liquid phase is called headspace gas chromatography. This technique is used in the cosmetics industry, in a variety of environmental applications, and in forensics. We will use it to measure the vapor pressures of two-liquid mixtures, as a function of the relative amounts of liquids present in the mixtures. Three liquid pairs will be investigated: acetone/chloroform; hexane/heptane; and acetone/heptane. Density and molecular weight data for these liquids are given in Table 1.

Table 1

Liquid	ρ (g/mL)	MW (g/mole)
acetone	0.79	58.08
chloroform	1.483	119.38
hexane	0.6603	86.18
heptane	0.6837	100.21

The equilibrium vapor pressure of a pure liquid is a measure of the "escaping tendency" of molecules from the liquid phase. The escaping tendency is determined by <u>intermolecular forces</u> -- forces exerted by molecules on each other -- in the liquid. These forces are always attractive, and can be represented between two molecules of substance A as shown below:

$$A \rightarrow \leftarrow A$$

Equation 1 represents the relation between P_{vap}, escaping tendency, and intermolecular forces.

$$P_{vap} \propto \text{escaping tendency} \propto 1/F \tag{1}$$

Table 2 gives vapor pressure data for three liquids at 25 °C, and their boiling points.

Table 2

Liquid	P_{vap} at 25°C	bp, °C
acetone	400 torr	56.5
ether	537	34.6
water	23.8	100

Consider a mixture of two liquids, A and B. The vapor pressure of A, $P_{vap}(A)$, over the liquid depends on the intermolecular forces acting on molecules of A in the liquid phase. A molecule of A will be acted on by A and B molecules as shown below:

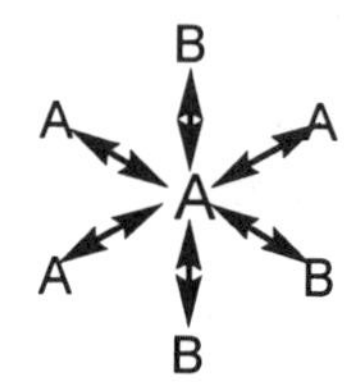

Suppose F(A-A) ≈ F(B-A). How will $P_{vap}(A)$ compare with the P^o_{vap} of pure A, in which an A molecule is surrounded only by A molecules? Equation 2 shows that it will depend on X_A, the fraction of molecules that

$$P_{vap}(A) = X_A\, P^o_{vap}(A) \qquad (2)$$

are A (X_A is also known as the mole fraction of A). If half the molecules are A, the vapor pressure of A will be 1/2 of what it would be if all the molecules were A. The same is true for substance B. Equation 2 is known as Raoult's Law. It is a statistical statement, valid only if all the various intermolecular forces are the same. Figure 1 shows plots of $P_{vap}(A)$, $P_{vap}(B)$, and P_{vap}(total) versus mole fraction of A, as predicted by

Figure 1
Ideal Solution

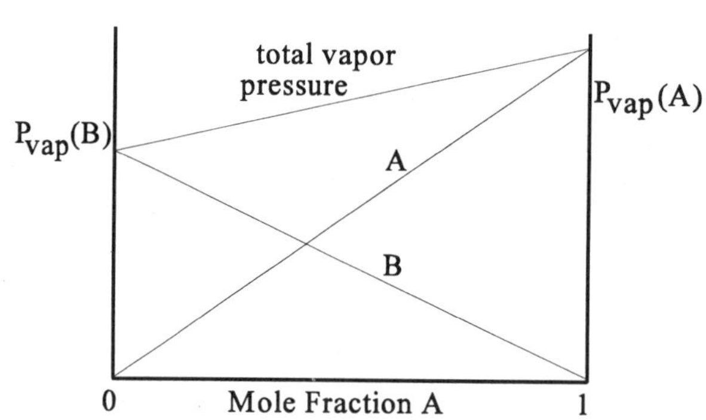

Raoult's Law. A solution in which the 2 components obey Raoult's Law is called an ideal solution.

Suppose now that F(A-B) > F(A-A), F(B-B). Now the escaping tendencies of A and B molecules from the solution will be less than from the pure liquids, so the vapor pressures are less than expected from Raoult's Law:

$$P_{vap}(A) < X_A P^o_{vap}(A) \qquad P_{vap}(B) < X_B P^o_{vap}(B)$$

The vapor pressures are said to exhibit negative deviations from Raoult's Law because they are less than Raoult's Law would predict. The total vapor pressure plot is shown in Figure 2.

Figure 2
Negative Deviations

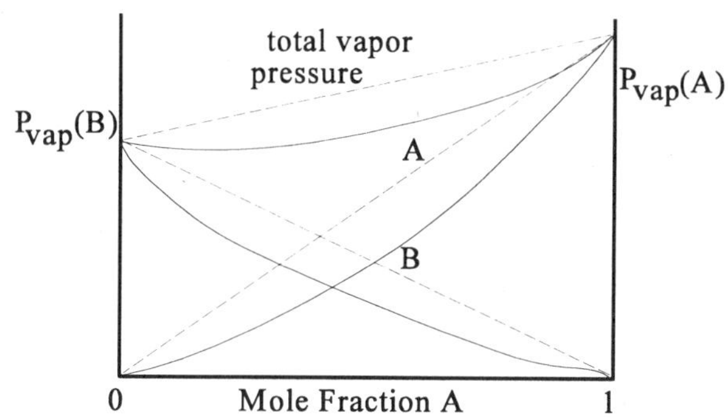

Finally, suppose that F(A-B) < F(A-A), F(B-B). This leads to positive deviations from Raoult's Law.

In this experiment you will examine one of the three liquid pairs acetone/chloroform, hexane/heptane, or acetone/heptane to answer the following question:

Does the liquid mixture conform to Raoult's Law?
Can this be understood in terms of the molecular structures of the liquids?

Equipment and Materials

gas chromatograph
2.0-mL gas tight syringes
1-mL graduated pipets
1-mL syringe pipet pumps
10-mL graduated cylinders
25-mL Erlenmeyer flasks with septum caps
reagents: acetone, chloroform, hexane, heptane

Safety

Eye protection must be worn at all times in the laboratory. All organic liquids used in this experiment are to be considered harmful. Avoid ingestion of liquid or inhalation of vapors.

MOST ORGANIC LIQUIDS ARE HIGHLY FLAMMABLE. KEEP AWAY FROM HEAT SOURCES AND OPEN FLAME.

Experimental

Record all data either in your lab notebook or on the **Results** sheet, whichever is applicable. Your section will investigate one of the three liquid pairs. Your group will be assigned one mixture of the two liquids, of different compositions. When your group has been assigned a mixture composition, set up three labelled 25-mL Erlenmeyer flasks. The two liquids, referred to below as A and B, should be apportioned into the flasks as follows:

Flask 1 $\approx$1 mL of pure liquid A;
Flask 2 Mixture of A and B, total volume 1.00 mL;
Flask 3 $\approx$1 mL of pure liquid B.

In a fume hood, measure out about 5 mL of each liquid into a CLEAN 10-mL graduated cylinder. From the cylinders, pipet the required volume(s) of liquid(s) into each flask and quickly fit the flask with a septum cap. (Do not leave a flask open to the atmosphere for any longer than necessary, as the liquids are volatile and will evaporate quickly.) Swirl each flask a few times to insure liquid-vapor equilibration. You are now ready to do headspace analysis.

The procedure for sampling the headspace vapor follows. It is important to do this in exactly the same way each time. Insert the gas-tight syringe through the center of the Erlenmeyer septum and slowly draw 0.2 mL of vapor into the syringe. Withdraw the syringe and draw in a further 0.6 mL of air (volumes of headspace vapor and air may be modified by your instructor). Insert the syringe into the GC port and slowly and smoothly eject the gas from the syringe into the GC. Begin with one of the pure liquids and make repeated injections during a single run to establish the proper attenuation factor and to satisfy yourself that you get good reproducibility in signal area. Repeat for the second pure liquid. Then repeat for your mixture, again making several injections during the run. Record all signal area data. Your data will be combined with data obtained by others, so exercise good experimental technique in obtaining it. Data will be analyzed during discussion.

Disposal Methods

When you have finished collecting data, DISPOSE OF YOUR LIQUID IN THE APPROPRIATE SOLVENT WASTE BOTTLE IN THE HOOD.

Questions

While waiting for the postlab discussion, work on the following questions.
1. Does your liquid pair obey Raoult's Law? If not, how does it deviate?
2. Which statement is true for your pair? $\Delta H_{soln} < 0$; $\Delta H_{soln} = 0$; $\Delta H_{soln} > 0$. Why?
3. Which statement is true for your pair? $\Delta V_{soln} < 0$; $\Delta V_{soln} = 0$; $\Delta V_{soln} > 0$. Why?

Preparation
Intermolecular Forces

Name __________________
Section __________________
Date __________________

Read

1. This experiment.
2. The appropriate sections of your textbook.

Problems

1. For a pair of liquids, A and B, which deviate positively from Raoult's Law, describe how the escaping tendencies of A and B molecules from the solution compare with what they are in the pure liquids. Describe how the vapor pressures of A and B compare with the values predicted by Raoult's Law.

2. Consider a mixture of two liquids, A and B. Assuming Raoult's Law to be valid, sketch plots showing how each of the following should vary as a function of the mole fraction of A in the mixture. The relative vapor pressure for a substance is the ratio of the observed vapor pressure for a substance to the vapor pressure of pure substance: P_{vap}/P^o_{vap}.

   a. The relative vapor pressure of A over the mixture
   b. The relative vapor pressure of B over the mixture
   c. The total relative vapor pressure over the mixture

3. Consider a mixture of liquids A and B, with $X_A = X_B = 0.5$. Which of the following statements is true, assuming $P^o_A > P^o_B$.

   a. The mole fraction of A in the vapor phase = 0.5.
   b. The mole fraction of A in the vapor phase > 0.5.
   c. The mole fraction of A in the vapor phase < 0.5.

Write a paragraph to explain your answer.

Results
Intermolecular Forces

Name ______________
Section ______________
Coworkers ______________
____
____
____
Date ______________

Data

Liquid pair:	liquid A =	________;	liquid B =	________
	Density =	________;	Density =	________
	MW =	________;	MW =	________

Mixture	Vol A	Vol B	10^3 x Moles A	10^3 x Moles B	Volume Fraction A	Mole Fraction A
1						

Signal Area (Vapor Pressure) Data

Flask #	Vol Fract A	Inj	Signal Area A	Signal Area B	Relative Area A	Relative Area B	Relative Area Total
1	1	1					
		2					
		3					
		4					
		5					
		6					
		Mean:	____	_____	1	0	1
		Attn:	_____				
2		1					
		2					
		3					
		4					
		5					
		6					
		Mean:	____	_____			
		Attn:	_____				

3	0	1				
		2				
		3				
		4				
		5				
		6				
		Mean: ____	____	0	1	1
		Attn: ____				

Conclusions based on section results:

Answers to Questions

Ideal Solutions

1 lab period; work in groups. Complete the Preparation page before laboratory.

Goals

- To gain experience in gas chromatography
- To discover the relationship between the vapor pressure of a pure liquid and its vapor pressure over a solution of a nonvolatile solute in the liquid

Background

Gas chromatography is a simple, fast, and accurate way to analyze volatile liquids. It can be used similarly to analyze mixtures of gases, the difference being that larger sample volumes must be injected to get observable signals. In this experiment, we will use gas chromatography to measure the vapor pressures over pure liquids and over solutions of solids in liquids in closed containers, to explore the effect of the solid solute on the vapor pressure of the liquid. Gas chromatography of a sample of vapor from the headspace over a liquid phase is called headspace gas chromatography. This technique is used in the cosmetics industry, in a variety of environmental applications, and in forensics. We will use it to measure the vapor pressure of a solvent over various solutions. Specifically, we will examine solutions of the solids, benzil, p-dichlorobenzene, and benzophenone in the solvents acetone, o-dichlorobenzene, and toluene. Molecular weight data for these substances and densities for the liquids are given in Table 1.

Table 1

Solid		MW (g/mole)
benzil		210.25
p-dichlorobenzene		147.01
benzophenone		182.21
Liquid	**ρ (g/mL)**	**MW**
acetone	0.79	58.08
o-dichlorobenzene	1.3048	147.01
toluene	0.8669	92.15

The equilibrium vapor pressure of a pure liquid is a measure of the "escaping tendency" of molecules from the liquid phase. The escaping tendency is determined by intermolecular forces exerted by molecules on each other in the liquid. These forces are always attractive, and can be represented between two molecules of substance A as shown below:

$$A\rightarrow\leftarrow A$$

Equation 1 expresses the relation between P_{vap}, escaping tendency, and forces.

$$P_{vap} \propto \text{escaping tendency} \propto 1/F \qquad (1)$$

Table 2 gives vapor pressure data for 3 liquids at 25 °C, and their boiling points.

Table 2

Liquid	P^o_{vap} at 25 °C	bp, °C
acetone	400 torr	56.5
ether	537	34.6
water	23.8	100

Consider a solution of a nonvolatile solute, S, in a solvent, A. The vapor pressure of A, $P_{vap}(A)$, over the solution depends on the intermolecular forces acting on molecules of A in the liquid phase. A molecule of A will be acted on by A and S molecules as shown below:

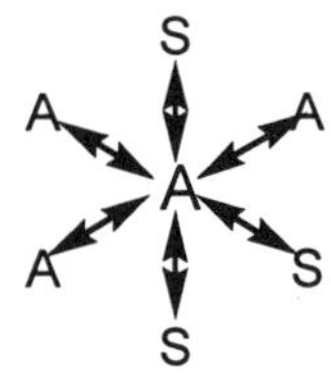

Suppose F(A-A) ≈ F(S-A). How will $P_{vap}(A)$ compare with the P^o_{vap} of pure A, in which an A molecule is surrounded only by A molecules? Equation 2 shows that it depends on X_A, the fraction of molecules that

$$P_{vap}(A) = X_A \, P^o_{vap}(A) \qquad (2)$$

are A (X_A is also known as the mole fraction of A). If half the molecules are A, the vapor pressure of A is 1/2 of what it would be if all the molecules were A. Equation 2 is known as Raoult's Law. It is a statistical statement, valid only if all the various intermolecular forces are the same. Raoult's Law predicts that plots of $P_{vap}(A)$ versus mole fraction of A should look as in Figure 1. A solution in which the vapor pressure of solvent obeys Raoult's Law is called an ideal solution.

Figure 1

Ideal Solution

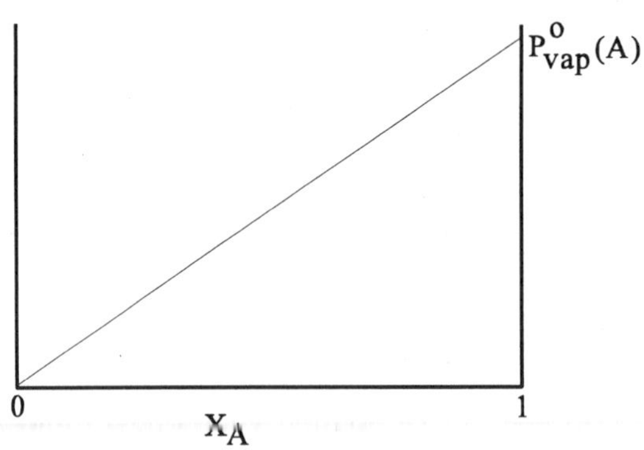

In this experiment you will examine the vapor pressure of a pure liquid and the vapor pressure over various solutions of a non-volatile solute in the liquid, to answer the following questions:

Does the solvent conform to Raoult's Law?
Can this be understood in terms of the molecular structures of the solute and solvent?

Equipment and Materials

gas chromatograph
2.0-mL gas tight syringes
5-mL serological pipets (graduated)
10-mL syringe pipet pumps
10-mL graduated cylinders
25-mL Erlenmeyer flasks with septum caps
reagents: acetone, o-dichlorobenzene, toluene, benzil, p-dichlorobenzene, benzophenone

Safety

Safety glasses must be worn at all times in the laboratory. All organic reagents used in this experiment are to be considered harmful. Avoid ingestion of solid or liquid or inhalation of vapors.

MOST ORGANIC LIQUIDS ARE HIGHLY FLAMMABLE. KEEP AWAY FROM HEAT SOURCES AND OPEN FLAME.

Experimental

Record all data either in your lab notebook or on the **Results** sheet, whichever is applicable. Your group will investigate one solute/solvent pair. Set up a series of 6 labeled 25-mL Erlenmeyer flasks. In one flask, place 2 mL of pure solvent. In the remaining 5 flasks, place 5 mL solvent and amounts of solute to give mole fractions of solute in the range 0.05 to 0.30. Use GC to determine the vapor pressure of pure solvent and the solvent vapor pressure over each solution. Your instructor will show you sampling and injection techniques.

Disposal Methods

If your solutions contain o-dichlorobenzene or p-dichlorobenzene, dispose of them in the waste bottle marked "chlorinated solvents." If your solutions contain neither of these two materials, dispose of them in the waste bottle marked "non-chlorinated solvents."

Questions

1. Does your solute/solvent combination obey Raoult's Law?
2. If not, what conclusions can you draw about S-A intermolecular forces?
3. Can you interpret your results in terms of molecular structures of solute and solvent?

Preparation
Ideal Solutions

Name ________________
Section ________________
Date ________________

Read

1. This experiment
2. The appropriate sections of your textbook.

Problems

1. What is the ordering of the strengths of the intermolecular forces in the liquids in Table 2?

2. The following vapor pressure data is obtained for solutions of a solute, S, in a solvent, A:

P_{vap}, torr	X_S, dimensionless
87.0	0 (pure solvent)
81.9	0.10
76.3	0.20
70.4	0.30
65.1	0.40

Does this solute/solvent pair obey Raoult's Law?

Results
Ideal Solutions

Name ____________________
Coworkers ____________________
____
____
____
____
Section ____________________
Date ____________________

Data

Solvent ____________________________ Solute______________________________

Solution	mass solute	volume solvent	X_S	X_A	Signal area	Relative P_{vap}
1	0					
2						
3						
4						
5						
6						

Conclusions:

Answers to Questions

Kinetics of Chemical Reactions

1 lab period; work in pairs. Complete the Preparation page before laboratory.

Goals

- To determine the overall rate law for a reaction
- To find the value of the rate constant

Background

The rate of a chemical reaction depends on the frequency with which the reacting molecules collide with the correct amount of energy. Frequency of collision is dependent on the concentrations of the reacting species, the number of molecules in a fixed volume. Molecular energy varies with the temperature. Thus, one would expect that by varying the concentration of one or more of the reactants, or the temperature of the reaction, the rate of the chemical reaction would vary. In this experiment, the temperature remains constant (room temperature) and the effect of concentration on the reaction rate is studied.

One of two reactions will be studied by your group and each involves the disappearance of the color of a solution over time. You may have noticed the color of a solution titrated to a phenolphthalein end point fades with time. Indicators, such as phenolphthalein (P^{2-}) and bromophenol blue (BP^{2-}), lose their color in basic solution. Fading is due to structural changes that occur when excess OH^- attacks the molecules as in reactions 1 and 2.

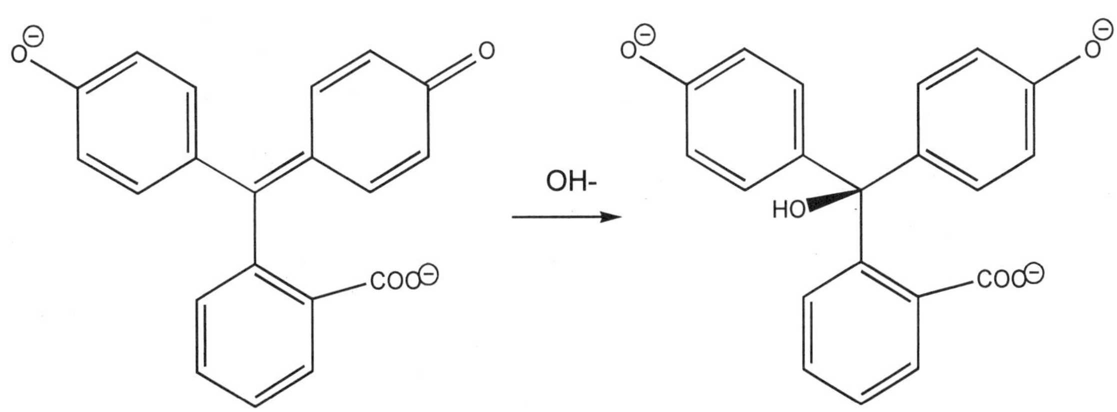

Reaction 1: Phenolphthalein and hydroxide ion

We can write the overall rate equation describing reactions 1 and 2, equation 1:

$$\text{rate} = k\,[IN]^m\,[OH^-]^n \qquad (1)$$

Here [IN] is the concentration of the indicator studied, $[OH^-]$ is the concentration of hydroxide ions, k is the overall rate constant, m is the order with respect to indicator and n is the order with respect to OH^-. To determine the rate law, the unknowns k, m, and n must be found. In the rate expression, the exponents are not the same as the coefficients of the balanced reaction. These values must be found by experiment.

Before beginning this type of study, you will need to have a method to follow the concentration of at least one reactant. A convenient way to follow changes is by monitoring the absorbance since A is directly proportional to the concentration of the colored species via Beer's law, equation 2:

$$A = \epsilon l C \qquad (2)$$

where ϵ is the molar absorptivity, l is the pathlength through the cell and C is the concentration of the species of interest. Since molar absorptivity is constant for a given molecule at a specific wavelength and the pathlength is fixed by the cell, absorbance, A, will only vary with C, the concentration of the absorbing species. Therefore, A may be substituted for concentration in the data analysis.

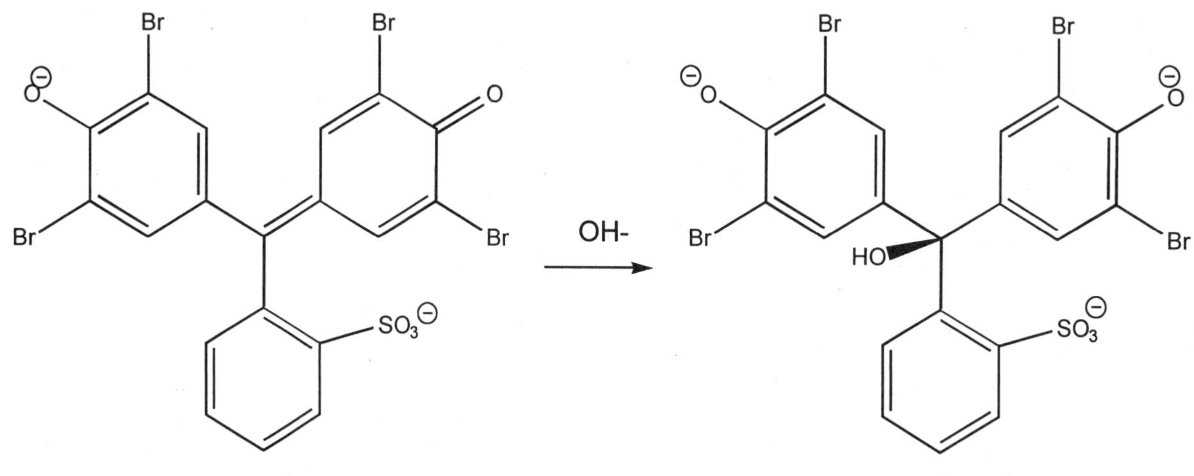

Reaction 2: Bromophenol blue and hydroxide ion

The order of a chemical reaction is found either by measuring the initial rate of the reaction or by monitoring the concentration of the reactant(s) or product(s) with time. If the initial rate method is used, the initial concentrations of the reactants are varied and the changes in initial rate are compared to the changes in the initial concentration. This method is often difficult to use. However, if the change in reactant concentration is followed over time and is plotted as a function of time the shape of the resulting curve is used to predict the order of the reaction with respect to the reactant followed. For example, for a reaction that is first order in species X, a plot of ln (X) vs. time is linear and the slope of the line is the negative of the rate constant. If a reaction is second order, a plot of 1/X against time is linear. Again, the slope is proportional to the rate constant. By obtaining a plot of the change in the concentration vs. time, the order of the reaction with respect to the colored species is known.

One method used to simplify determining the reaction order with respect to one species is to set up conditions so that the other reactant is in large excess. Thus, the concentration of that reactant is considered constant during the course of the experiment. For a reaction in which one reactant is colored so that its concentration may be easily followed with a spectrometer, and the second reactant is in large excess so that its concentration is essentially constant, the analysis of the data becomes significantly simplified.

Your group will be assigned an indicator to study. Since you will be comparing data with others in the class, you will need to have a constant concentration of indicator and of NaCl for all of the reactions studies.

Are the overall rate constants for the fading of indicators the same?

Equipment and Materials

100 mL volumetric flasks
stop watch
spectrometer, Spectronic 20 or other
10 mL Mohr pipets or 1 to 10 mL volumetric pipets
4 M NaOH
4 M NaCl
bromophenol blue 0.5 g/L
phenolphthalein 1 g/L

Experimental

Carry out the reactions as discussed in the prelab briefing. Record results in your notebook or on the **Results** page as instructed. When adding the base to the indicator, begin timing the reaction as soon as the solutions mix, even before completely diluting to volume. Once the cuvette is filled, it may be left in the spectrometer until the experiment is complete. Record the absorbance at regular intervals. Plot the data using a computer graphing program and determine the order with respect to the indicator. Next, determine if the concentration of OH^- is part of the rate law. If OH^- is part of the rate law, what is the order with respect to OH^-? Wait for the class discussion. If time permits, prepare a solution without NaCl. In this new solution, keep the concentration of indicator and NaOH the same as the solution studied above. Does changing the solution background change the value of the rate constant measured? Explain.

Disposal Methods

Solutions may be washed down the drain with excess water unless otherwise directed by the instructor.

Questions

1. What is the value of the rate constant obtained for your reaction conditions?
2. What is the order of the reaction with respect to the indicator?
3. Does the concentration of OH^- affect the rate of the reaction?
4. What is the overall order of your reaction?
5. Write the overall rate equation.
6. Turn in all concentration-time plots.

References

1. "Kinetics of the Fading of Phenolphthalein in Alkaline Solution," by Lois Nicholson. *Journal of Chemical Education,* 1989, 66, 725-726.
2. "Fading of Bromophenol Blue," by Randall Winans and Charles Allen Brown. *Journal of Chemical Education,* 1975, 52, 526-527.

Preparation
Kinetics of Chemical Reactions

Name ___________
Section___________
Date ___________

Read

1. This experiment.
2. About kinetics and Beer's law in your text.

Questions

Define the following terms:
rate constant
order of a reaction
pseudo order reactions

Practice Problem

Calculate the amount of each stock reagent needed to prepare 100.0 mL of the following solutions given the new concentrations:

stock	new concentration	mL of stock needed
0.5 g/L BP	0.005 g/L	___________
1 g/L P	0.01 g/L	___________
5 M NaCl	3 M NaCl	___________
5 M NaCl	0.3 M NaCl	___________
4 M NaOH	2 M NaOH	___________
4 M NaOH	0.5 M NaOH	___________

Describe how you would prepare a solution that is 0.005 g/L in BP^{2-} and 2 M NaOH from the available chemicals. Show calculations for the amounts used.

Describe how you would prepare a solution that is 0.005 g/L in BP^{2-} and 1 M NaOH from the available chemicals and 2 M NaCl. Show calculations for the amounts used.

Results
Kinetics of Chemical Reactions

Name ____________
Coworker(s) ____________
____
Section ____________
Date ____________

Data

Indicator studied__________
Concentration of indicator in final solution_________
Concentration of NaCl in final solution___________

Table 1. Time and Absorbance data

Trial 1		Trial 2	
Final conc. NaOH________			________
Time	Absorbance	Time	Absorbance

Answers to Questions

Results
Kinetics of Chemical Reactions

Name ______________
Coworker(s) ______________
____
Section ______________
Date ______________

Data

Optional study of the effect of inert salt on the rate of reaction.

Indicator studied__________

Concentration of indicator in final solution_________
Concentration of NaOH in final solution___________

Table 2. Time and Absorbance data

Trial 1		Trial 2	
Final conc NaOH___________		___________	
Time	Absorbance	Time	Absorbance

Answers to Questions

NMR

1 lab period; work in groups. Complete the Preparation page before laboratory.

Goals

- To learn the basic principles of nuclear magnetic resonance spectroscopy
- To determine the structure of a molecule from its NMR spectrum

Background

The nuclei of some atoms, but not all, behave as if they were rotating, or spinning, about an axis passed through them, much as a top spins about its central axis. Examples of atoms whose nuclei spin are hydrogen, fluorine, the 13 isotope of carbon, and the 15 isotope of nitrogen. There are many others not listed here. Under certain conditions, this spinning motion, which is of course an oscillatory motion, can interact with and absorb energy from electromagnetic radiation in the radio frequency region of the spectrum. The nuclear spin is a motion of relatively low energy, so low energy electromagnetic radiation is required to interact with it. A typical radio frequency photon has a frequency of 10^8 s^{-1}, a wavelength of 300 meters (about the length of three football fields), and an energy of 6.6×10^{-26} Joules. By measuring the energies of photons absorbed by the spinning nuclei, we obtain the nuclear magnetic resonance (NMR) spectrum of the molecule. This spectrum provides information about the chemical environment of the spinning nucleus, and can be used to deduce the atomic bonding patterns in the molecule. Sophisticated NMR techniques are used to determine the 3-dimensional structures of huge protein molecules in solution. Recently, nuclear magnetic resonance spectroscopy has found medical applications, in which context it is known as magnetic resonance imaging (MRI). MRI uses the spinning motions of atomic nuclei to provide a map of the internal structure of human body tissue.

The nuclear magnetic resonance spectrum resulting from spin of the hydrogen atoms in the molecule acetaldehyde is shown in Figure 1. The structure of acetaldehyde is also shown. Several aspects of this spectrum are important. First, the single sharp signal at exactly 0 ppm is not due to acetaldehyde, but is instead due to a reference compound named tetramethylsilane, or TMS. The structure of TMS is also shown in Figure 1. By general agreement, the positions of NMR signals are always reported relative to the position of the signal for TMS. Second, the unit used to measure resonance position, ppm (parts per million), is not a unit of energy. It has no units at all. The ppm unit is a measure of the shift in position between the TMS resonance and a particular resonance in the compound of interest. It is usually called the chemical shift. The chemical shift gives an indication of the chemical environment in which the spinning nucleus is found. Third, there are two groups of signals, or resonances, due to acetaldehyde. These are centered at 2.2 ppm and 10 ppm in Figure 1. Note that one group consists of two sharp lines of equal size (i.e., intensity). The total intensity of this pair of sharp lines is 24 (obtained from the integration, which measures the area under the signal). Taken together, the pair of lines is called a doublet. The 10-ppm signal consists of four equally spaced lines which, taken together, are called a quartet, with total intensity of 8. Study this spectrum and try to make some sense of it in terms of the acetaldehyde structure. Discussion of this system in lab will enable us to discover some general rules about NMR spectra.

In this experiment, you will be given three compounds for which the formulas are known but the arrangement of atoms is not. Your goal will be to answer the following questions for each compound:

What arrangements are possible for the atoms, consistent with the formula?

Which arrangement is correct, based on the NMR spectrum?

Figure 1

Acetaldehyde

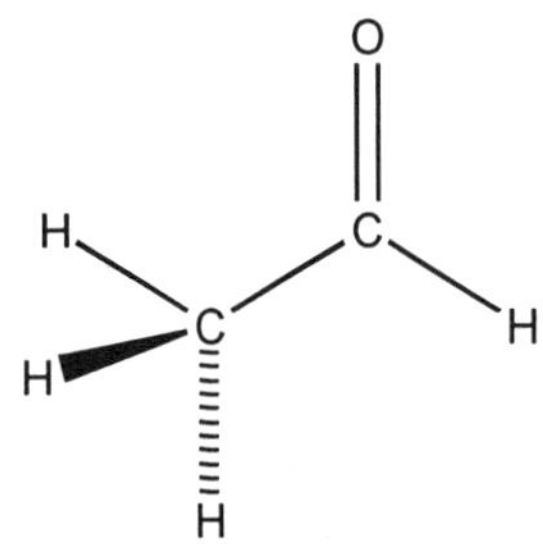

Tetramethylsilane

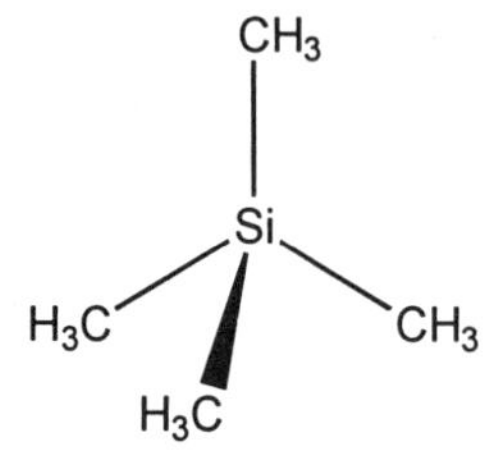

NMR Spectrum of Acetaldehyde

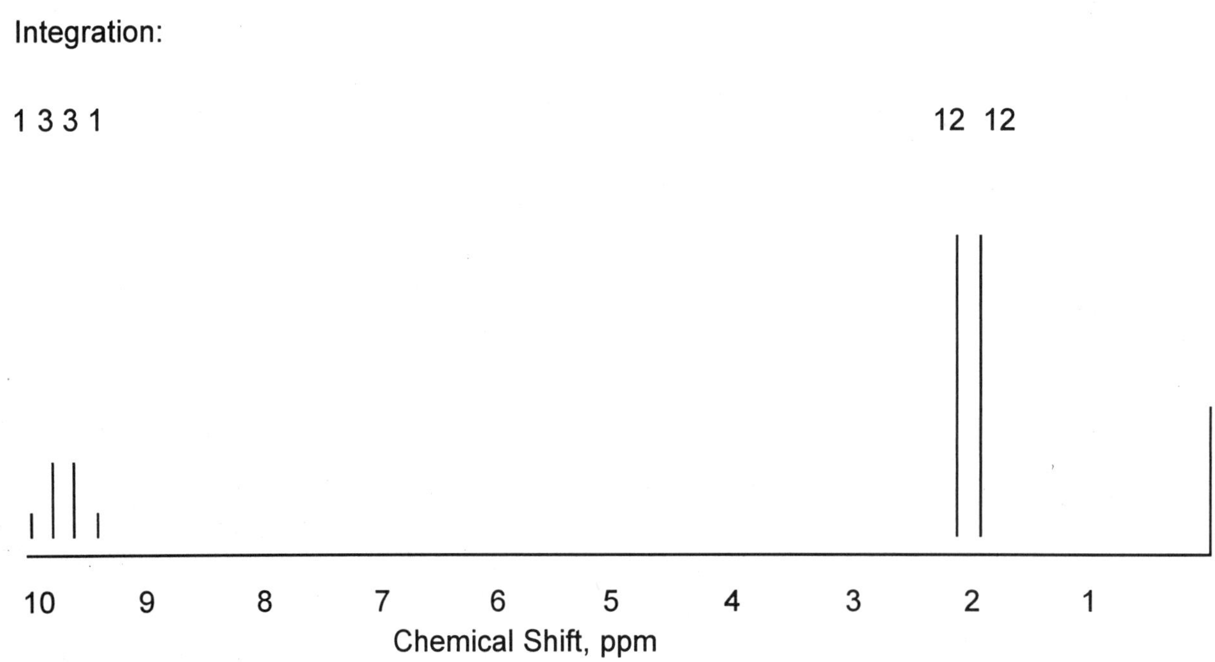

Equipment and Materials

NMR spectrometer
NMR tubes
deuterochloroform solutions of several unknown compounds having formula:

$C_4H_{10}O$
C_3H_6O
C_3H_7Cl
$C_3H_6Cl_2$

Safety

Safety glasses must be worn at all times in the laboratory. Deuterochloroform is hazardous. If you break an NMR tube, tell your instructor immediately so that the solution may be cleaned up.

Experimental

Record all data in your lab notebook or on the **Results** sheet, whichever is applicable. Obtain three NMR tubes from the instructor. Each contains a deuterochloroform solution of a compound of known formula but unspecified structure (arrangement of atoms). First, for each compound, draw as many structures as you can consistent with the formula. Work together. Remember that carbon atoms form 4 bonds, nitrogen atoms 3, oxygen atoms 2, halogen atoms 1, and hydrogen atoms 1. When satisfied that you have drawn all possible structures, predict the appearance of the NMR spectrum for each structure drawn. Predictions should address the number of signals (i.e., the number of distinct types of hydrogen atoms in the structure) and the splitting of each signal due to interaction with spins of hydrogens on neighboring atoms. Predict the relative chemical shifts of the various types of protons, based on electronegativity arguments.

Second, obtain the NMR spectra of the samples using the NMR spectrometer. When you are finished obtaining spectra, return all sample tubes to the instructor.

Disposal Methods

You are not required to do any disposal. Samples will be disposed of by course personnel at the completion of the experiment.

Questions

1. A compound has the formula C_3H_5N. What are the possible structures consistent with the formula? Describe the NMR spectrum for each structure.

Preparation
NMR

Name ____________________
Section ____________________
Date ____________________

Read

1. This experiment
2. The appropriate sections of your text.

Problems

1. The structure of propionaldehyde, C_3H_6O, is shown in Figure 2. Predict what the NMR spectrum of propionaldehyde should look like.

Figure 2

Propionaldehyde

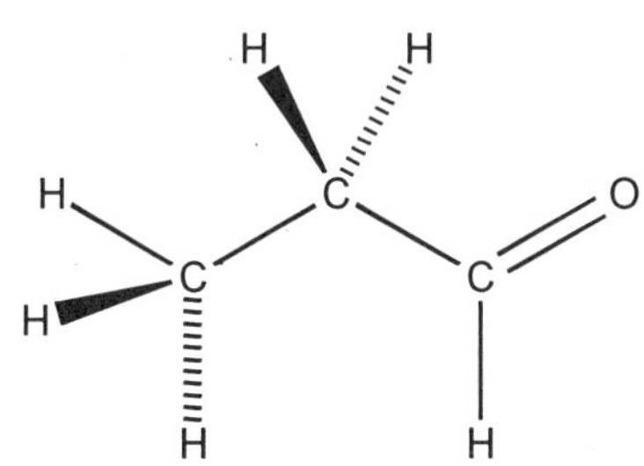

Results
NMR

Name ____________________
Section ____________________
Coworkers ____________________
____
____
____
Date ____________________

Data

Numbers of Compounds: ________ ________ ________

Formulas of Compounds: ________ ________ ________

Possible Structures:

Compound ____ Compound ____ Compound ____

Correct structures, based on NMR:

Answers to Questions

Workshops

Graphing

(Developed with LH Berka)

1 lab period; work individually. Complete the Preparation page before laboratory.

Goals

- To learn
  a. The purpose and advantages of graphical analysis of data
  b. Some of the functional forms that occur commonly in chemistry
  c. How to cast a functional relationship into linear form
  d. The meanings of the terms dependent variable, independent variable, Cartesian coordinate system, origin, slope, intercept, x-axis, y-axis, abscissa, ordinate, interval size, subdivisions of the graph paper, numerical limits
  e. How to properly construct a graph, whether by hand or by computer, including
    1. Correct choice of the quantity to be plotted on each axis
    2. Good choice of numerical limits for each axis, to bracket the experimental points and allow for easy interpolation of readings between interval marks on the axes
    3. Correct plotting of experimental points, with each point marked so it can be seen
    4. Correct method for obtaining the best straight line or smooth curve through the data points
    5. Method for determining the numerical value and units for the slope and intercept of a linear plot
- To learn the use of a spreadsheet for organizing, manipulating, and plotting data

Background

Scientists experiment. Normally, the goal of an experiment is to determine how the value of one variable, called the dependent variable, changes as the experimenter systematically alters the value of another variable, called the independent variable. For example, a chemist may be interested in how the volume, V, of a gas sample varies as its pressure, P, is changed; or in how the rate of a reaction changes with the concentration of one of the reactants. There are two common ways of presenting the results of experiments like these: tabular form and graphical form. Although both methods are commonly used, and are often used together, the graphical method is generally superior because it gives the scientist a visual feeling for the relationship between the dependent and independent variables.

Graphs are usually plotted on a Cartesian axis system as in Figure 1. The vertical axis is called the y-axis (or ordinate) and the horizontal axis, the x-axis (or abscissa). Generally, the value of the dependent variable (or some function of it) is plotted on the y-axis, and that of the independent variable (or some function of it) on the x-axis. We assume that you have some familiarity with the Cartesian coordinate system and know how to plot a point represented by the ordered pair (x,y) in such an axis system.

A plot of the dependent variable, y, versus the independent variable, x, can be either linear or curved. For example, a plot of temperature in °C versus the corresponding temperature in °F is linear (straight line). On the other hand, a plot of the solubility of a salt such as $K_2Cr_2O_7$ (potassium dichromate) versus temperature is curved. Although both linear and curved plots can be useful, the linear plot is generally more so, for several reasons. First, linearity or deviations from it are readily seen by eye. On the other hand one cannot tell visually if a curve that resembles a hyperbola is truly hyperbolic. For example, although the

Figure 1

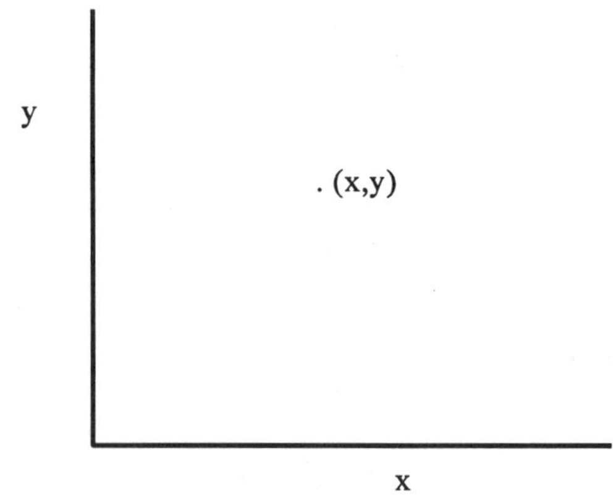

Figure 2
Hyperbolic or Not?

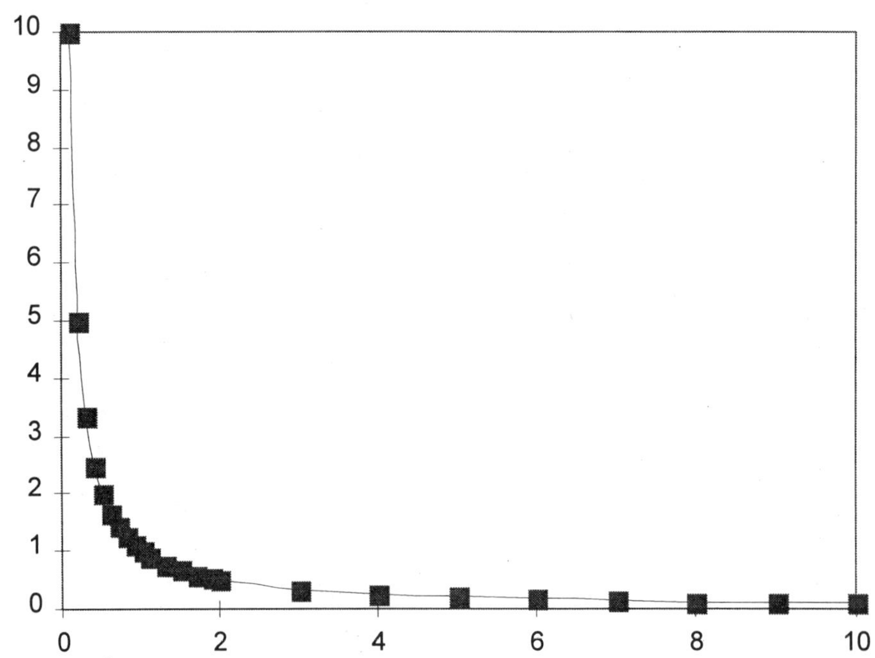

curve in Figure 2 looks hyperbolic, it is not obvious from inspection that it actually is. Second, linear plots may be readily extrapolated (extended) into regions which are experimentally not accessible. Extrapolation allows the prediction of the value of the dependent variable under conditions other than those governing the experiment. For these reasons, scientists prefer linear plots. This workshop will focus on these.

The equation governing a linear relationship between two variables, x and y, is given in equation 1, where y is the dependent variable, x is the independent variable, m is a constant called the slope, and b is a constant called the y-intercept.

$$y = mx + b \qquad (1)$$

These quantities are illustrated in Figure 3. The trick to constructing a linear plot is to cast the functional

Figure 3

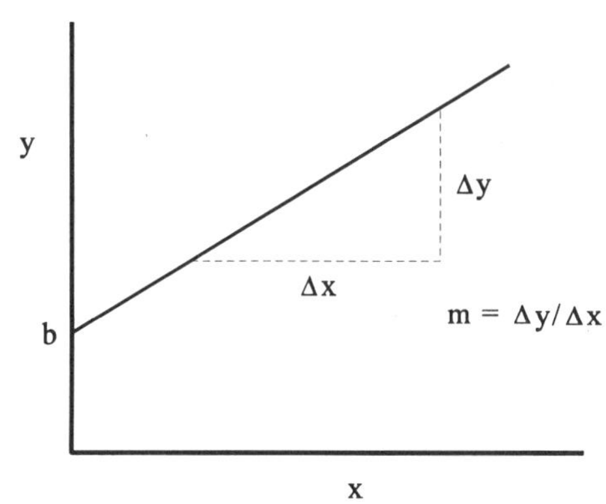

relationship between dependent and independent variables into the form of equation 1. For example, the relationship between the pressure, P, and volume, V, of a fixed amount of ideal gas at constant temperature is in equation 2.

$$PV = K \qquad (2)$$

where K = a constant. This equation does not have the form of equation 1. However, dividing both sides by P gives

$$V = K/P \qquad (3)$$

Equation 3 has the form of equation 1 where y corresponds to V, x to l/P, m to K, and b to 0. A plot of y (V) versus x (1/P) should be a straight line with slope m (K) and intercept zero; i.e., the line should pass through the origin.

Methods for Plotting Graphs

Graphs may be constructed by hand or by computer. A professional scientist or engineer is able to use both methods. The purpose of this workshop is to provide guidance and experience in both plotting methods. The general criteria for construction of a graph are the same regardless of method. Properly used, they will enable you to produce a plot that has a professional appearance and is easy to read and interpret.

Common Functional Forms in Chemistry

If we agree to use y to represent the dependent variable (the variable measured in the experiment) and x to represent the independent variable (the variable controlled by the experimenter), there are six functional relationships between y and x which occur often in chemistry. These are detailed below. In the following discussion, do not be concerned if you do not understand where the equations come from and what the quantities mean. For now, we are interested only in the forms of the equations and how these translate into plots.

Case 1. $y = mx + b$. The value of y varies linearly with the value of x.

The volume of a sample of gas at constant pressure is related in this way to the temperature in °C. Illustrative data are shown in Table 1.

Table 1

Independent Variable, T (°C)	Dependent variable, V (L)
0	0.403
30	0.450
60	0.485
80	0.515
100	0.560
150	0.625

Clearly V increases as T increases, so a plot of V versus T might be linear. Figure 4 shows such a plot, created using a computer plotting software package. If we put an eye down near the paper and sight along the points, it is clear that they fall along a straight line. The computer has calculated and drawn the straight line which best fits the data. Note that as many points as possible lie near the line, with roughly an equal number above and below it. The line represents the functional relation between the dependent variable, V, and the independent variable, T. Slope and intercept, calculated by the computer using the method of linear regression analysis (or method of least squares), are shown in the diagram:

$$V = 0.001492\, T + 0.4019 \tag{4}$$

The relationship between volume and temperature for a gas is called Charles' Law. The units of the slope, calculated by dividing ΔV in liters by ΔT in °C are L/°C. Intercept units are liters. Several important features of Figure 4 are listed below.

1. The <u>entire window</u> (for a manual plot, the window is the sheet of graph paper) is used for the plot. Generally, the bigger the graph the better, subject to the constraints in 2.

Figure 4
Gas Volume versus Temperature

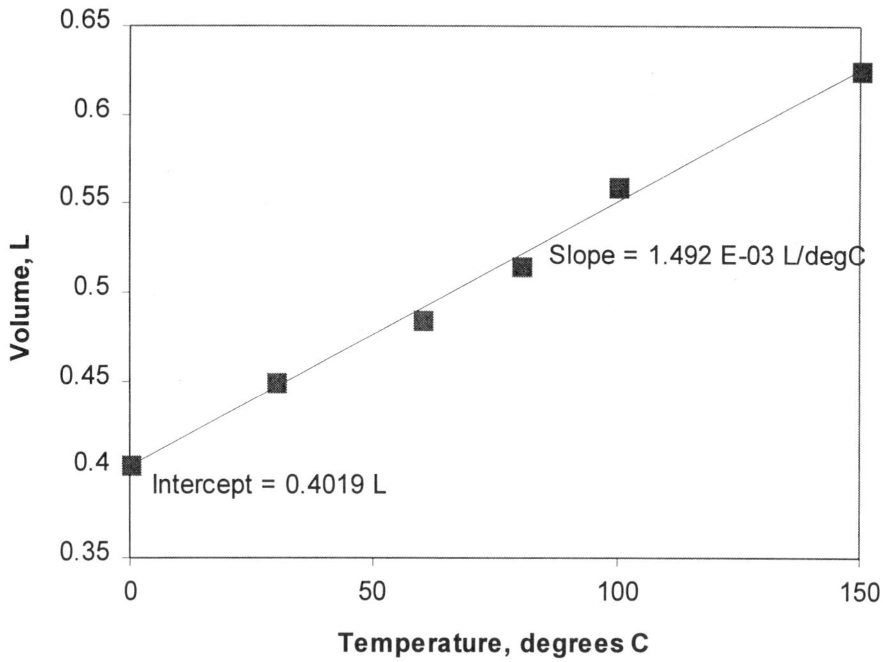

2. The <u>upper</u> and <u>lower</u> axis limits of the variable being plotted (0.65 L and 0.35 L for the y-axis in Figure 4) are chosen with two constraints.

a. The lower and upper limits must <u>bracket</u> the range of experimental values being plotted; i.e.,the lower limit must be ≤ the smallest experimental value of the variable, and the upper limit must be ≥ the largest experimental value of the variable. Thus the lower limit of the y-axis in Figure 4 must be < 0.403 and the upper limit > 0.625. Bracketing should be tight.

b. The correspondence between the number of units spanned between the lower and upper axis limits and the number of intervals (steps) along the axis must allow for convenient plotting and reading of points. This means that the number of variable units per basic interval must be 2, 2.5, 5, or 10. (These numbers are generic--depending on the order of magnitude of the numbers plotted, the decimal point could be anywhere--e.g., 0.2, 0.25, 0.5, and 1.0).

3. The lower limit of each axis need not be zero. Although zero is a convenient lower limit for the T axis in this case, choice of zero for the lower limit of V would cause all the points to be cramped into a small area of the graph, as in Figure 5. The requirement that the plot fill as much as possible of the window takes precedence.

4. Last, although some of the points miss the line (experimental error), we do <u>not</u> "connect the dots". Always put the best smooth curve or line through the points.

Constraint 2b is usually easier to satisfy when plotting is done by computer. For manual plotting, it is more difficult, because we are restricted by the physical layout of commercial graph paper, with <u>basic intervals</u> of either cm or inches. Each basic interval is further subdivided into 5 or 10 parts. The graph paper you

Figure 5
Gas Volume versus Temperature

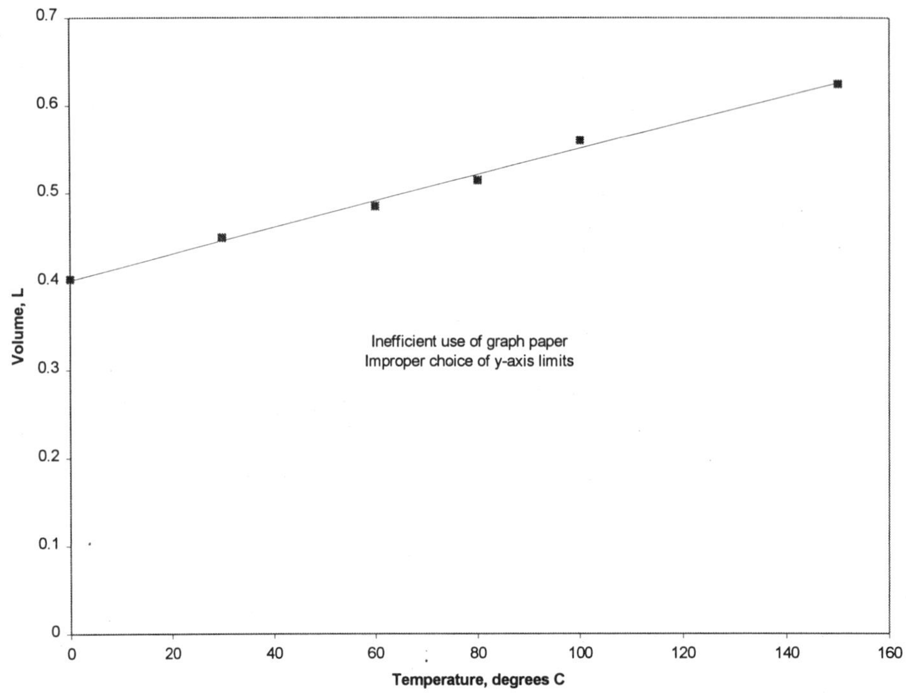

bought for this experiment has the cm as the basic interval with each subdivided into 10 parts. This paper is described as "10 squares (or subdivisions) to the centimeter." There are usually 24 cm along the long dimension of the paper and 18 cm along the short dimension. In plotting a graph, one must establish a correspondence between the number of units spanned by the variable being plotted and the number of basic intervals available along the axis. The values of t range from 0 to 150 °C. The lower x limit must be ≤ 0 and the upper limit $\geq$ 150. However, there are 18 basic intervals along the x-axis. To set up the x-axis so that the left end of the line corresponds to 0 and the right end of the line to 150 would require

$$(150 - 0)/18 = 8.33 \text{ °C per cm.}$$

This is an inconvenient number, so constraint 2b is not met. To satisfy both constraints, set the upper limit of the axis at 180 °C. This exceeds 150 and gives a convenient number of variable units per basic interval, specifically, 10 °C per cm. Setting the x-axis up to span the range 0 to 180°C is as close as we can come to using the whole length of the axis, while maintaining plotting convenience. Similarly, V spans the range between 0.403 and 0.625 L, and there are 24 basic intervals along the y-axis. The number of variable units spanned is 0.625 - 0.403 = 0.222. The ratio of this number to 24 is inconvenient. Therefore picking lower and upper limits such that a variable range of 0.240 L is spanned gives a convenient correspondence of 0.01 L per cm. The limits used for the y-axis in a manual plot would be set up this way. Finally, the straight line in the manual plot is usually drawn by eye. To calculate the slope in a hand-plot, choose two points which lie on the line, not two experimental points. The best straight line averages the data, so a slope calculated from two points exactly on the line is more accurate than a slope based on two experimental points. The two points chosen should lie at the upper and lower extremes of the plot, rather than near one another on the line.

Again, this is to minimize error in the slope due to error in reading y values from the line. Finally, we choose two points whose x values are convenient--that is, which correspond with lines delineating the basic intervals of the graph paper. Normally this procedure will give values close to the regression values produced in a computer plot.

Whether plotting by hand or computer, setting convenient limits becomes easy with practice, and is one of the goals of this workshop.

Case 2. $y = m/x + b$. y is inversely proportional to x.

A relationship of this type is that between the volume, V, of a sample of gas and its pressure, P, at constant temperature. Illustrative data is in Table 2.

Table 2

Dependent Variable: V, L	Independent Variable: P, atm	1/P
3.67	0.10	10
1.45	0.25	4.0
1.02	0.36	2.8
0.860	0.42	2.4
0.620	0.60	1.7
0.400	0.90	1.1
0.330	1.1	0.91

As P increases, V decreases, suggesting an inverse relation. A linear plot might be obtained by plotting V versus 1/P. Calculated values of 1/P are given in the 3rd column of Table 2, and a computer plot of V vs. 1/P is in Figure 6. Upper and lower variable limits of the axes have been chosen by the guidelines above. The plot is clearly linear. The slope, calculated by regression analysis, is 0.368 L-atm. The intercept is zero, within experimental error, so the equation relating the variables, called Boyle's Law is

$$V = 0.368 /P \qquad (5)$$

Case 3. $y = mx^n$, with n an integer. y is proportional to x raised to an integral power.

If n = 1, this equation reduces to case 1; if n = -1, it reduces to case 2 with b = 0. This type of functional relationship occurs frequently in chemical kinetics, where the rate of reaction is often proportional to an integral power of the concentration of a reactant. There are two steps to the graphical analysis of data related by a function of this form.

a. A plot of ln y vs. ln x is constructed. The reason for this is seen by taking the natural logarithm of both sides of the equation, $y = mx^n$:

$$\ln y = \ln (mx^n) = \ln m + n\ln x \qquad (6)$$

A plot of ln y versus ln x should be linear, with slope n.

Figure 6

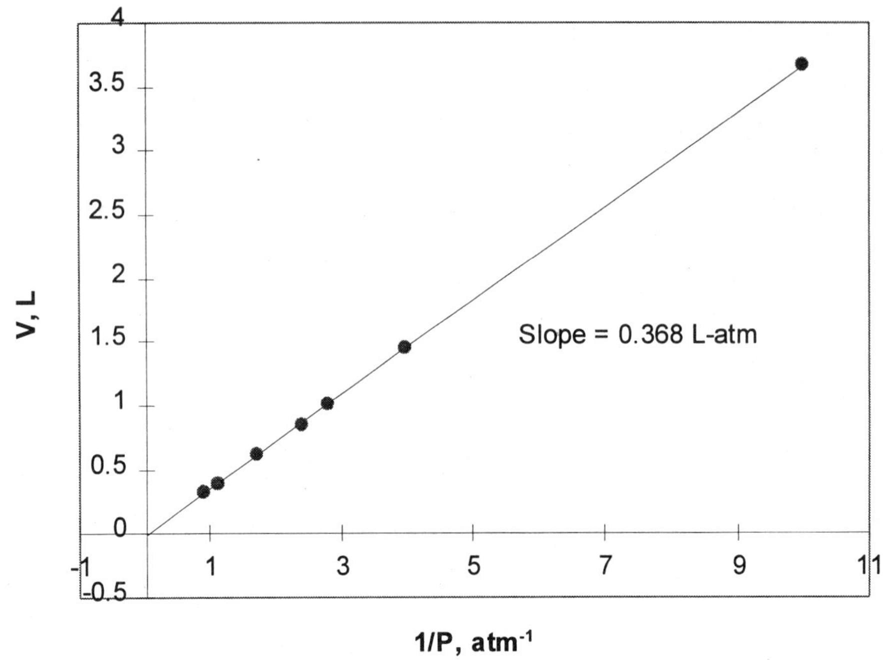

(There are two systems of logarithms in common use: logarithm to the base 10 (abbreviated $\log_{10}$); and natural logarithm (abbreviated ln). Natural logarithm is also called logarithm to the base e, where e is the number 2.71828.... These logarithms are defined as follows:

$\log_{10} x$ is a number y such that $10^y = x$.

ln x is a number z such that $e^z = x$.

It can be shown that the relationship between the two logarithmic scales is simple:

$\ln x = 2.303 \log_{10} x$)

Once n has been determined,

b. y is plotted against x^n. m may be determined from the slope.

Table 3 shows the variation of the rate of the reaction in equation 7 as a function of the concentration of NOCl in moles per liter.

$$2NOCl \rightarrow 2NO + Cl_2 \qquad (7)$$

Table 3

Dependent variable Rate, moles/L-sec	Independent Variable [NOCl], moles/L	$[NOCl]^2$
3.60×10^{-9}	0.30	0.09
1.44×10^{-8}	0.60	0.36
3.24×10^{-8}	0.90	0.81
4.80×10^{-8}	1.10	1.21
7.40×10^{-8}	1.35	1.82
11.4×10^{-8}	1.70	2.89

Assuming this data obeys $y = mx^n$, we first plot ln y vs. ln x to obtain n. Values of ln (Rate) and ln [NOCl] are in Table 4.

Table 4

ln (Rate)	ln [NOCl]
-19.442	-1.204
-18.056	-.511
-17.245	-.105
-16.852	0.0953
-16.419	0.300
-15.987	0.531

A computer plot of ln (Rate) vs. ln [NOCl] is shown in Figure 7. The regression slope is 1.997, very close to 2, so n = 2. We now return to the original data and plot Rate vs. $[NOCl]^2$ to obtain m from the slope.

Values of $[NOCl]^2$ are in column 3 of Table 3, and a plot of Rate versus $[NOCl]^2$ is in Figure 8. The slope of this plot is 3.96×10^{-8} L/mole-sec. The original data obey equation 8.

$$\text{Rate} = 3.96 \times 10^{-8}[NOCl]^2 \qquad (8)$$

Before proceeding, we should note something about the graph in Figure 8. The numerical values of rate (Table 8) are very small, involving an exponential factor of 10^{-8}. It is clumsy to write such small numbers along the axis of a graph. To avoid this, scientists label the axis using only the <u>argument</u> of the number (e.g., the argument of the number, 4.80×10^{-8} is 4.80). The exponential term is acknowledged in the axis label. This has been done in Figure 8, where the vertical axis is labeled, Rate (moles/L-sec) x 10^8. The number, 10^8, is the number by which all the experimental rates were multiplied to convert them to the argument, which is actually plotted. Thus

$(4.80 \times 10^{-8}) \times 10^8 \rightarrow$ 4.80
argument - this is the number plotted.

Figure 7

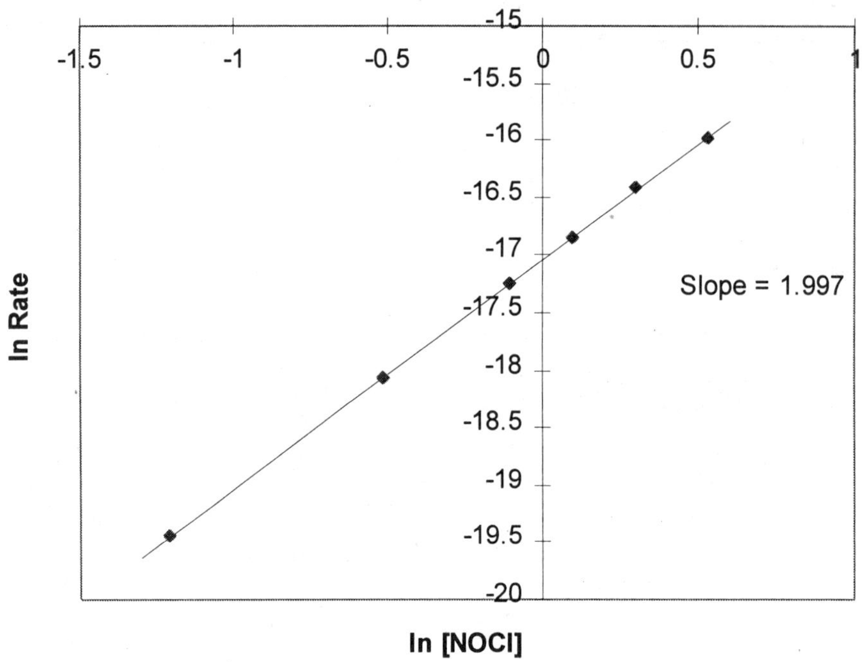

Figure 8

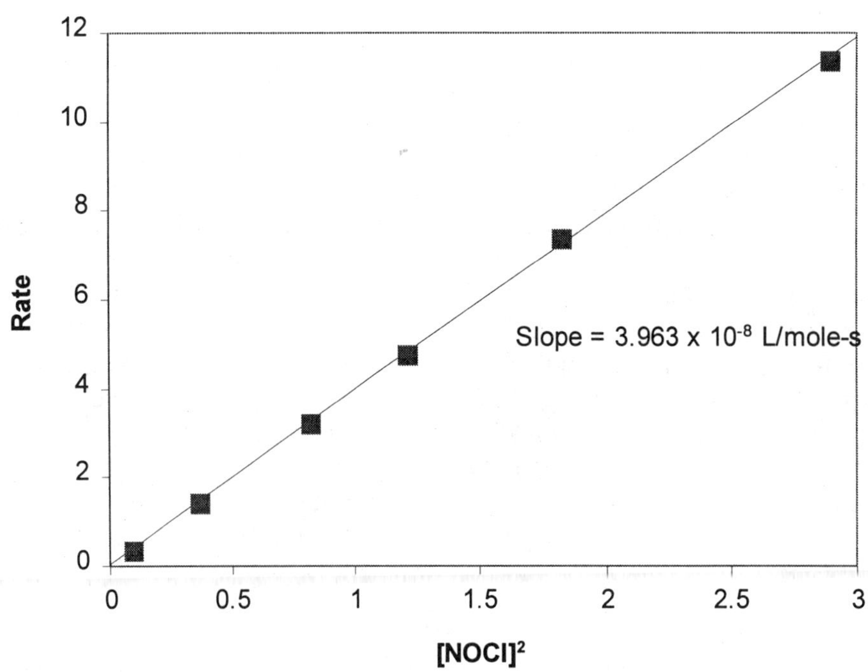

For plots in which the numbers on one or both axes have implied exponentials, allow for these in calculating slope and intercept manually. Please see your instructor with questions about this.

Case 4. $1/y = mx + b$.

Again, this form is common in kinetics. For a second-order reaction, the concentration of the reactant varies with time according to equation 9, where t = time elapsed since the beginning of reaction, k = the second-order rate constant, and $[\text{reactant}]_o$ is the concentration of reactant at t=0, the beginning of the reaction.

$$1/[\text{reactant}] = kt + 1/[\text{reactant}]_o \qquad (9)$$

A specific example of a second-order reaction is in equation 10. Table 5 shows how the concentration of C_4H_6 (butadiene) varies with time.

$$2C_4H_6(g) \rightarrow C_8H_{12}(g) \qquad (10)$$

The third column lists calculated values of $1/[C_4H_6]$, which are plotted against time in Figure 9. The plot is linear, with slope 0.0141 L/mole-s and intercept 59 L/mole. From the intercept, $[C_4H_6]_o$ = 1/intercept = 0.0170 M.

$$1/[C_4H_6] = 0.0141\ t + 59 \qquad (11)$$

Table 5

t, sec	$[C_4H_6]$, moles/L	$1/[C_4H_6]$, L/mole
195	1.62×10^{-2}	61.7
604	1.47×10^{-2}	68.0
1246	1.29×10^{-2}	77.5
2180	1.10×10^{-2}	90.9
4140	0.89×10^{-2}	112
4655	0.80×10^{-2}	125
6210	0.68×10^{-2}	147
8135	0.57×10^{-2}	175

Case 5. $\ln y = mx + b$.

Kinetics again provides examples of this type of function. For first-order reactions, concentration of reactant varies with time according to equation 12.

$$\ln [\text{reactant}] = -kt + \ln [\text{reactant}]_o \qquad (12)$$

Figure 9

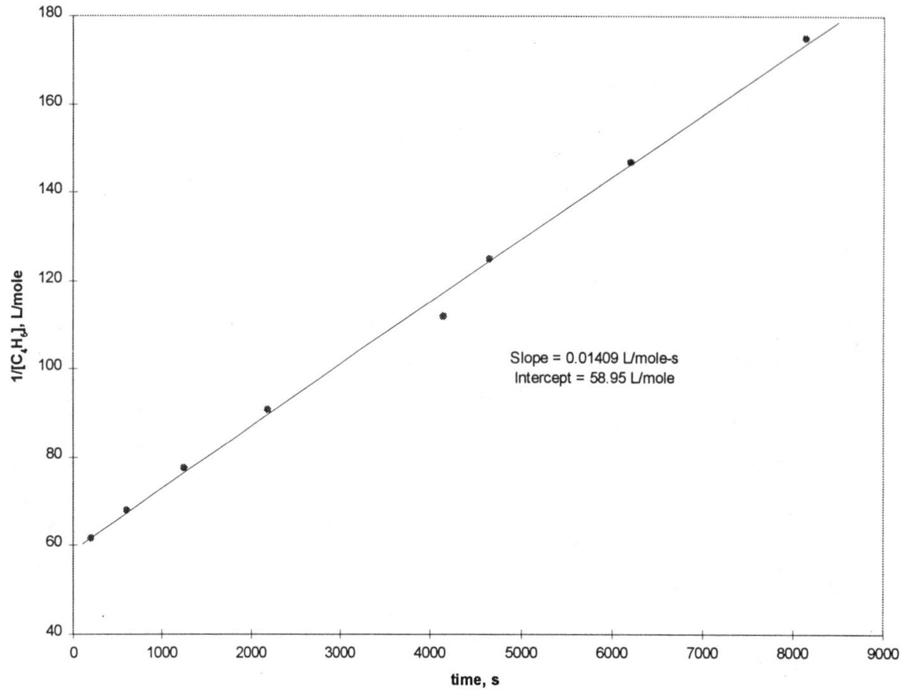

where t = time elapsed since the beginning of reaction, k is the first-order rate constant for the reaction, and $[\text{reactant}]_o$ is the concentration of reactant at t=0.

An example of a first-order reaction is eqn 13.

$$2N_2O_5(g) \rightarrow 4NO_2(g) + O_2(g) \tag{13}$$

Concentration-time data are in Table 6.

Table 6

$[N_2O_5]$, mole/L	t,sec	ln $[N_2O_5]$
5.00	0	1.609
3.52	500	1.258
2.48	1000	0.908
1.75	1500	0.560
1.23	2000	0.207
0.87	2500	-0.139
0.61	3000	-0.494

Values of ln $[N_2O_5]$ are in the third column of the table, and a plot of ln $[N_2O_5]$ vs. t is in Figure 10. The plot is linear. The rate constant, k (slope), is 7.0×10^{-4} sec^{-1} (be sure you understand how units are determined). $\ln [N_2O_5]_o$ (intercept) is 1.609.

Figure 10

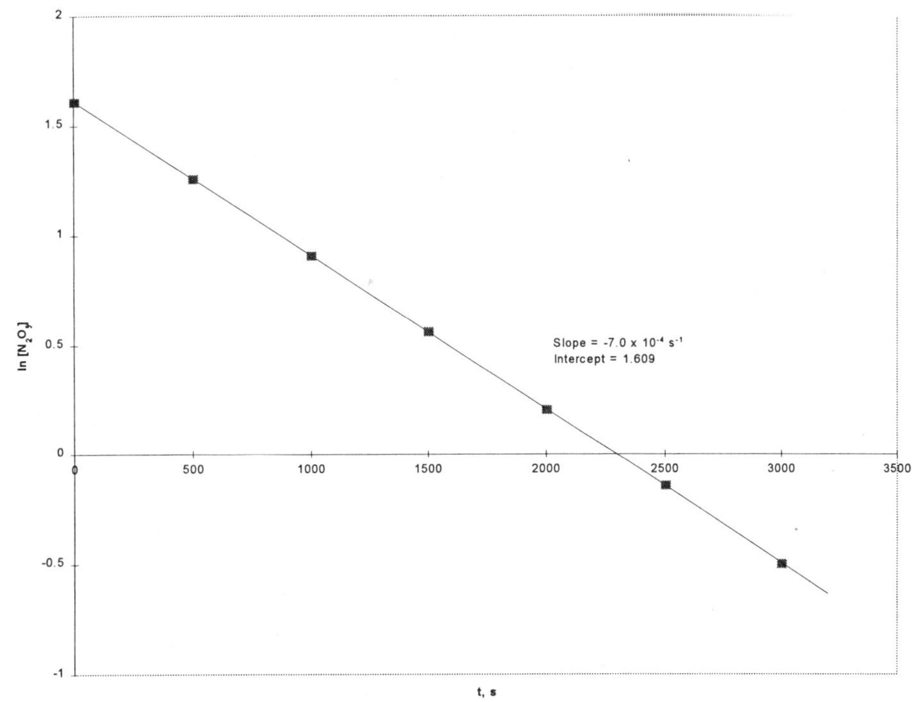

Case 6. This is a very common functional relation in chemistry.

$$\ln y = m/x + b \qquad (14)$$

Some examples include how the equilibrium constant, K_{eq}, for a reaction, the rate constant for a reaction, and the vapor pressure of a pure liquid vary with temperature. Consider a particular example.

For most chemical reactions, the rate constant varies with Kelvin temperature according to eqn 15.

$$\ln k = -E_a/RT + \ln A \qquad (15)$$

Here k is the rate constant, T is the temperature, E_a is the activation energy for the reaction, A is the frequency factor for the reaction, and R is the gas constant. (Do not worry about where this equation comes from or what all of the quantities mean--focus on the functional form and the graphical analysis).

Table 7 shows the rate constant as a function of temperature for reaction 16.

$$H_2 + I_2 \rightarrow 2HI \qquad (16)$$

Table 7

Temperature (°C)	k (L/mole-sec)
283	1.2×10^{-4}
302	3.5×10^{-4}
355	6.8×10^{-3}
393	3.8×10^{-2}
430	1.7×10^{-1}

To plot ln k versus 1/T (K), we must a) convert temperature to the Kelvin scale; b) reciprocate the Kelvin temperatures; c) find the natural logarithms of the rate constants. The results are in Table 8.

Table 8

T(K)	1/T, K^{-1}	ln k (no units)
556	1.799×10^{-3}	-9.028
575	1.739×10^{-3}	-7.958
628	1.592×10^{-3}	-5.573
666	1.502×10^{-3}	-3.270
703	1.422×10^{-3}	-1.772

A plot of ln k vs. 1/T is in Figure 11. Slope and intercept are

slope $= -1.93 \times 10^4$ K
intercept = 25.57

Since slope = $-E_a/R$, and R = 8.314 J/mole-K, we calculate E_a as follows

$$E_a = (-8.314 \text{ J/mole-K}) \times (-1.93 \times 10^4 \text{K})$$

$$= 1.60 \times 10^5 \text{ J/mole}$$

Since intercept = ln A, we calculate

$$A = e^{\text{intercept}} = e^{25.57} = 1.27 \times 10^{11}$$

In all of the previous examples, zero was a convenient lower limit for the range of the independent variable. In this example, however, zero is not a convenient lower limit for the abscissa if we want to utilize the full dimensions of the window. Since the zero of the abscissa does not occur where the two axes meet, the y-intercept cannot be obtained manually by extrapolating the linear plot to intersect the y-axis. In cases like this, the value of the y-intercept must be calculated. The standard form for a straight line, equation 1, can

Figure 11

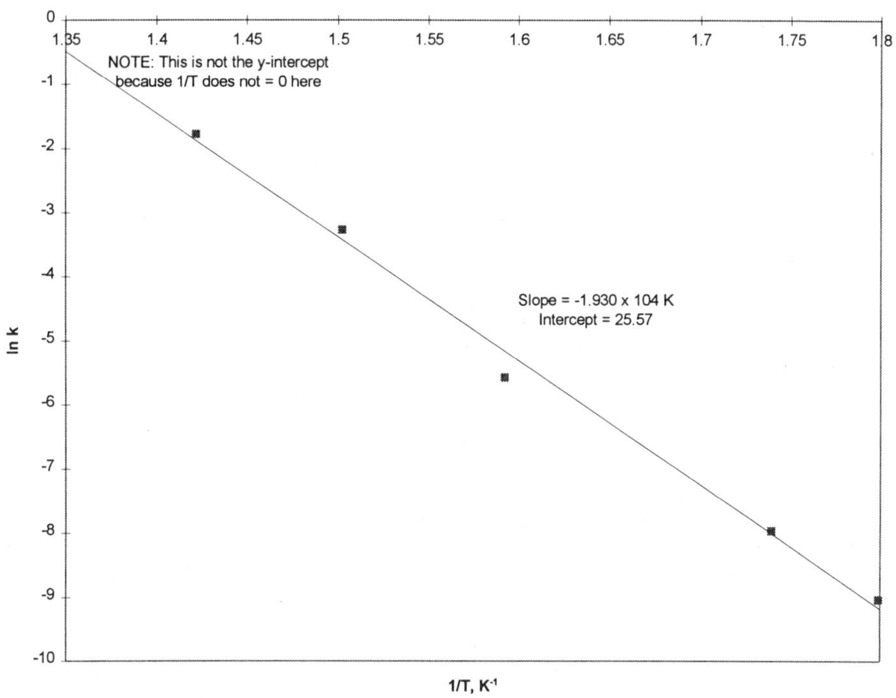

be rearranged to equation 17.

$$b = y - mx \tag{17}$$

First, the slope is calculated from the best-fit line for the data. Then a single point on the line is chosen, and b is calculated by 17.

The intercept in Figure 11 may be calculated this way.

$$\ln A = \ln k + E_a/RT$$

Choosing the point for which ln k = -7.23 and $1/T = 1.7 \times 10^{-3}$, we obtain

$$\ln A = (-7.23) - (-1.93 \times 10^{4})\ (1.7 \times 10^{-3})$$

$$= 25.6$$

This completes our survey of common linear functional forms. Although other forms occur, familiarity with those discussed here will serve you well.

Before concluding this section, we summarize the practical aspects of constructing and analyzing plots.

1. For hand plots, use regulation graph paper, preferably ruled with 10 subdivisions per centimeter.
2. Decide how to plot the experimental data. If possible, plan a linear plot. How you plot the data (e.g., as y vs. x, y vs. 1/x, ln y vs. 1/x, etc.) depends on the expected relationship between the variables.
3. Perform calculations consistent with your chosen plotting method. For example, if you decide to plot the data as ln y vs. 1/x, calculate ln y and 1/x for each data point.
4. For a hand plot, set up a sheet of graph paper:
    a. Darken the coordinate axes using a straight edge. Use the left vertical line as the y-axis, and the bottom horizontal line as the x-axis.
    b. Label the axes with the quantities to be plotted.
    c. Decide on the numerical limits (lowest and highest values) to appear on the axis for each variable.
        1. The limits should <u>bracket</u> the experimental values
        2. The limits should be chosen so there is a convenient number of <u>variable units</u> per <u>basic interval</u> of the graph paper.
        3. It is not necessary to start with zero on each axis. In many cases this causes compression of the data points into too small a region of the graph.
        4. Within constraints 4c 1 and 2, use as much of the window (graph paper) as possible for the plot.
5. Plot the points according to the chosen format (step 3). In a hand plot, circle each point to indicate its location, and draw the best straight line or smooth curve through the points. Use a straight edge or french curve.
    a. <u>Never</u> "connect the dots." Not all points fall exactly on the straight line or smooth curve, since each point has experimental error. Deviation of a point from the best straight line or curve gives a measure of the error in the point.
6. If the graph is linear, determine the slope and intercept. For a computer plot, use linear regression analysis. For a hand plot:
    a. Slope $=\Delta y/\Delta x$. Use two <u>widely separated</u> points <u>on the line</u>.
    b. Intercept. If the lower limit of the abscissa is zero, extrapolate (extend) the straight line to intersect the y-axis. The y value at the intersection is the intercept. If the lower limit of the abscissa is not zero, the intercept must be calculated: $b = y - mx$. Choose any point (x,y) on the line. Use x, y, and m to calculate b.

Required Materials

Bring the following items to lab:

2 pieces of graph paper, preferably K & E type, with 10 subdivisions per centimeter.
a straight edge
several sheets of notebook paper
one electronic calculator
2 sharpened pencils
an appropriate blank floppy disk

Experimental

You will be given five sets of experimental data. During the laboratory period, construct, by hand, both required plots for the first of the five data sets (Experiment #1, the temperature dependence of vapor pressure), and decide which of the two functional forms best fits the data. From the plot of the best functional form, determine the <u>value</u> and <u>units</u> of slope and intercept, and the values of any other specified quantities. Show graphs, slope, and intercept to your instructor. S/he will point out errors, which must be corrected before leaving the lab.

After your two hand plots are approved, obtain a copy of a spreadsheet program on your blank floppy disk by submitting your disk to your instructor. Instructions for using the spreadsheet will be provided. During the week following your lab period, use the spreadsheet to produce the nine plots required and to do the calculations. If you need help with spreadsheet, see your instructor. Turn in the plots, calculations, and completed **Results** sheet at your next lab meeting. CAUTION: The spreadsheet will calculate values, but not units, of slope and intercept, and it will give many more significant figures than are appropriate. It is necessary for you to determine units and to report values to an appropriate number of significant figures on the **Results** sheet.

Disposal Methods

No disposal required.

Preparation
Graphing

Name ________________

Section ________________

Date ________________

Read

1. This experiment.
2. The appropriate sections of your textbook.

Problems

1. The following data show the pressure exerted by a gas as a function of its volume:

Volume, mL	Pressure, torr
33.3	600
20.0	1000
14.3	1400
12.5	1600
10.0	2000

a. Plot P (on the ordinate) versus V (on the abscissa).

b. Make a second plot to obtain a straight line. From the graph, obtain value and units of

1. The slope
2. The intercept

2. You measure a quantity, y, as a function of a quantity, x, and find the following functional relation between them:

$$1/y = 1/\alpha x^3 + \beta/\alpha$$

The units of y are moles L^{-1} s^{-1} and those of x are moles L^{-1}. The slope of a plot of 1/y vs. $1/x^3$ is 10.12 and the intercept is 2.84. From this information determine

a. units of the slope

b. units of the intercept

c. numerical value and units of α

d. numerical value and units of β

Results
Graphing

Name ______________
Section ______________
Date ______________

Experimental Data, Set A

Experiment 1:

Correct functional form ______________________________
Slope =
Intercept =
γ =
β =

Experiment 2:

Correct functional form ______________________________
Slope =
Intercept =
α =
δ =

Experiment 3:

n =
γ =

Experiment 4:

Correct functional form ______________________________
α =
β =

Experiment 5:

Valid Functional Relationships:
Constants:

Results
Graphing

Name ______________
Section ______________
Date ______________

Experimental Data, Set B

Experiment 1:

Correct functional form ______________________________
Slope =
Intercept =
γ =
β =

Experiment 2:

Correct functional form ______________________________
Slope =
Intercept =
k =
θ =

Experiment 3:

Correct functional form ______________________________
α =
β =

Experiment 4:
n =
γ =

Experiment 5:

Valid Functional Relationships:
Constants:

Spreadsheets

1 week; to be completed outside of laboratory.

Goals

- To become familiar with a spreadsheet program
- To plot data using the graphing routine of a spreadsheet program

Background

Spreadsheets are commercially available software packages designed for the fast and convenient handling of information in tabular form. Most widely used sheets incorporate a number of features that facilitate not only the organization of information, but also allow manipulation and plotting of data in a variety of ways. It is probably not an exaggeration to say that the modern professional or pre-professional must add the spreadsheet to his or her arsenal of data handling techniques in order to be competitive in our fast-paced modern world. There are a number of spreadsheets that are widely used and readily available, including among others Lotus-123®, Microsoft Works®, Microsoft Excel®, and QuattroPro®. Most spreadsheets incorporate essentially the same array of operations which are accessed in much the same way. Therefore it is possible to present "generic" directions for spreadsheet use, which can then be made specific for a particular spreadsheet. Since your instructor most likely has a favorite spreadsheet that s/he would like you to use, we will not attempt here to give detailed instructions for using any particular one. Instead, we give general instructions which can be expanded upon by your instructor. The instructions given here are tailored to the preceding workshop in the manual, **Graphing**. By applying the instructions to the vapor pressure-temperature data from the graphing workshop, you will become familiar with the basic operations of the spreadsheet. You should then be equipped to apply the spreadsheet to any desired data set.

Your instructor will assist you in obtaining access to a spreadsheet program. To successfully carry through the instructions below, you need a blank, formatted disk which can be used to store worksheets and plots that you create using the spreadsheet.

Loading the Spreadsheet

Access the spreadsheet from the computer hard drive or a working disk, as instructed.

Creating a Worksheet Containing Data to be Plotted

Refer to the reproduction of the finished worksheet in Figure 1 during the following discussion.

Enter a title for the worksheet in cell A1. For example, "Graphical Analysis of Data, Experiment 1"

Move the pointer to the first cell in the third row (this is done with the arrow keys or the mouse). In many spreadsheets, this cell is called A3.

Type a label for the first column of numbers in the set of data points. For example, "P_{vap},atm".

Type a label for the second column of numbers in cell B3. For example, "T(K)".

Move the pointer to cell A5.

Enter the first data point in the first two columns. Put the first number in the first cell in the row, the second in the second cell in the row. In Figure 1, the first point has 0.0062 under P_{vap}, atm, and 273 under T(K).

Enter the second data point in the next row. Continue until all data points have been entered (for Experiment 1, you should finish in row 15.)

Performing Calculations on the Data in Order to Make Plots

For Experiment 1, you must calculate ln P_{vap} and 1/T for each data point (you may or may not be able to subscript portions of row or column labels in the spreadsheet that you use. Subscripts are shown in this

Figure 1

	A	B	C	D	E	F	G	H	I
1	Graphical Analysis of Data, Set A, Experiment 1								
2									
3	Pvap, atm	T(K)	lnPvap	1/T	lnPvap(cal	lnPvap(calc)			
4									
5	0.0062	273	-5.08321	0.003663		-5.09145			
6	0.0136	285	-4.29769	0.003509		-4.28839			
7	0.0241	294	-3.72554	0.003401		-3.72911			
8	0.034	300	-3.38139	0.003333		-3.37491			
9	0.059	310	-2.83022	0.003226		-2.81503			
10	0.097	319	-2.33304	0.003135		-2.34115			
11	0.174	331	-1.7487	0.003021		-1.7494			
12	0.268	340	-1.31677	0.002941		-1.333			
13	0.506	355	-0.68122	0.002817		-0.68592			
14	0.732	364	-0.31197	0.002747		-0.32327			
15	1	373	0	0.002681		0.021883			
16									
17									
18	SUMMARY OUTPUT								
19									
20	*Regression Statistics*								
21	Multiple R	0.999975							
22	R Square	0.999951							
23	Adjusted	0.999945							
24	Standard	0.012464							
25	Observati	11							
26									
27	ANOVA								
28		*df*	*SS*	*MS*	*F*	*gnificance F*			
29	Regressio	1	28.33751	28.33751	182410.5	1.08E-20			
30	Residual	9	0.001398	0.000155					
31	Total	10	28.3389						
32									
33		*Coefficient*	*andard Err*	*t Stat*	*P-value*	*ower 95%*	*pper 95%*	*ower 95.0*	*pper 95.0*
34	Intercept	13.98129	0.038393	364.1668	4.52E-20	13.89444	14.06814	13.89444	14.06814
35	X Variable	-5206.86	12.19132	-427.095	1.08E-20	-5234.44	-5179.28	-5234.44	-5179.28

description for clarity.) The spreadsheet will do these calculations for you. Place the ln P_{vap} values in column C, and 1/T in column D.

Enter a label in cell C3; e.g., ln P_{vap}.
Do the same in cell D3; e.g., 1/T
In cell C5, enter the formula for calculating the natural log of the vapor pressure in cell A5. Your instructor will explain how to do this. Copy the contents of cell C5 to rows 6-15 of column C.
In cell D5, enter the formula for calculating the reciprocal of the temperature in cell B5. Copy the contents of D5 to rows 6-15 of column D.

Performing Linear Regression Analysis of the Data

Ln P_{vap} vs. T Plot

Your instructor will show you how to do a linear least squares analysis with the spreadsheet. Carry out a linear regression by specifying the T values in B5-B15 as the x-range, and the ln P_{vap} values in C5-C15 as the y-range. Store the output at an appropriate location in the spreadsheet.

Ln P_{vap} vs. 1/T Plot

Repeat the regression analysis, this time using the 1/T values in column D as the x range. At this point the regression analyses for two possible fits of the data should be adjacent to each other in the spreadsheet. The correlation coefficient, R^2, gives a measure of how well the data fit the calculated straight line. The closer R^2 is to 1.0, the better the fit. Look at your two R^2 values. Which functional form best fits the data?

Plotting the Data

LnP$_{vap}$ vs. T Plot

Access the plotting routine from the main menu (this is often called Graphics) and follow your instructor's directions for creating a plot of ln P_{vap} versus T. Be sure to title the plot; scale the axes appropriately, so the graph is easy to read and interpret; and save the plot as a file.

Ln P_{vap} vs. 1/T Plot

Repeat the plotting procedure to obtain the second required plot. Give it a different name. Now you have both required plots incorporated as part of your worksheet. At this point you should quit the plotting submenu and return to the worksheet.

Saving the Worksheet and Plots

Insert your personal disk into the computer disk drive. Your instructor will show you how to save your worksheet and plots so that you can access them at any future time.

Printing a Plot

Print your plots by following the instructions specific to your spreadsheet.

Printing the Worksheet

If desired, print a hard copy of your worksheet.

Solids

(Developed with LH Berka)

1 lab period; work in pairs. Complete the Preparation page before laboratory.

Goals

- To understand several fundamental aspects of the structures of crystalline solids:
  a. Hexagonal closest packing (hcp) and cubic closest packing (ccp)
  b. Tetrahedral hole
  c. Octahedral hole
  d. Face-centered cubic (fcc) lattice
  e. The equivalence of ccp and fcc
  f. The meaning of the phrase, unit cell
  g. Locations of tetrahedral and octahedral holes in a fcc unit cell
  h. The crystal structures for which the following compounds, all based on a face-centered cubic unit cell, are prototypes
     1. NaCl
     2. Zinc-blende (ZnS)
     3. Fluorite (CaF_2)
     4. Anti-fluorite
     5. Diamond
  i. Simple cubic lattice:
     1. the CsCl-type structure
  j. Body-centered cubic (bcc) lattice
  k. Relationship between ionic radius and structure

Background

Sir W. H. Bragg, one of the first scientists to use X-rays to study crystal structures, was fascinated by the intricate, orderly atomic patterns revealed by his investigations. In 1901 he likened the stability of crystalline structures to that of an engineering structure.

Natural crystals, from gemstones to snowflakes, have always held a fascination for man. Today we recognize that the regular geometric shapes of crystals reflect orderly three-dimensional networks of the atoms, ions, or molecules that form the crystal lattice.

Although the arrangements of atoms and ions that lead to regular crystal lattice structures are quite simple, it is hard to visualize the arrangements in 3-dimensions. Two-dimensional drawings of 3-dimensional objects do not help! The purpose of this workshop is to give you the opportunity to build models of some common crystal structures and clearly see how the atoms and ions are arranged. Having built and examined models, it will be easier for you to make sense of drawings in the text and class discussions.

Perfect crystals, in which the repetitive lattice pattern is completely uniform throughout, are extremely rare. This is not surprising when you consider that a crystal is formed from a liquid (a solution or the molten solid) or a gas, in which particles are randomly oriented. To initiate crystal growth, a tiny collection of particles must form in a highly ordered pattern. This process, called nucleation, may occur spontaneously when particles of a pure substance collide with appropriate orientations and with low enough kinetic energies to allow "permanent" interparticle attractions rather than "rebound collisions." More often, a foreign material provides a surface on which particles may be adsorbed in orientations suitable for continued crystal growth. The nucleation of snowflakes, for example, often involves a "dust particle" as the center of adsorption for water molecules, and a string suspended in a supersaturated solution of a salt may provide nucleation sites as ions are adsorbed onto the fibers.

The tiny microcrystals initially formed may be perfectly ordered, but continued growth introduces imperfections. As the crystal grows, the surface area for addition of more particles becomes larger, requiring an ever-increasing availability of the proper particles with appropriate orientations. When molecular crystals grow from the pure liquid (e.g., as pure water freezes) or gas (e.g., in formation of iodine crystals from vapor), there are no problems with imperfections from foreign particles, but it is unlikely that uniform growth rates over all surface areas will continue. Nonuniform growth results in crystal imperfections. Crystals grown from mixtures, such as salt solutions or impure liquids, may contain foreign particles trapped in the lattice in addition to physical imperfections resulting from irregular growth patterns.

The growing of nearly perfect crystals, or of crystals containing "planned imperfections," is important in modern solid state technology. Synthetic diamonds are grown for such uses as diamond-tipped drills; sodium chloride crystals are produced for optics systems of infrared spectrometers; and "doped" crystals of silicon or germanium (containing planned substitutional lattice defects) are prepared for solid state electronic equipment.

Equipment and Materials

Crystal model kit, consisting of
- cabinet with 5 shelves, each with 25 holes
- a closest packing shelf (grooved)
- 90 red spheres
- 30 blue spheres
- 10 yellow spheres

Experimental

Write answers to questions in your notebook or on the **Results** page, as instructed. Each group should obtain a model kit from the instructor. As you do the experiment, make simple sketches of the models. Describe each structure using the shorthand notation for atom positions discussed in the pre-lab briefing. Answer questions as you go. Indicate which question you are answering by number.

1. Closest Packing
   a. Place the grooved closest-packing shelf on the top shelf of the kit with grooves running left to right.
   b. Make a close-packed layer of red spheres using the grooves, consisting of 7 rows of spheres. The number of spheres per row should alternate 9,8. . .9. In a close packed layer, the atoms of one row fit the indentations in the neighboring rows, as shown. This arrangement allows the largest possible number of spheres to fit in a given area--i.e., the density of spheres is maximum. Each sphere is surrounded by a hexagon of spheres touching it.

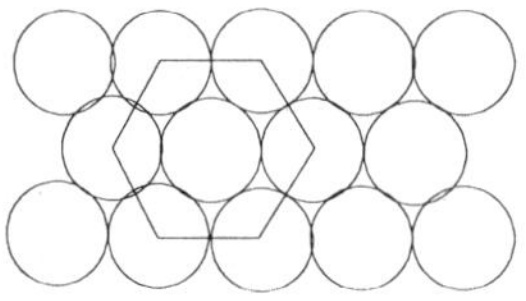

   c. View your close-packed layer from above. You will see two types of triangular holes between spheres, shown below. One type has a vertex pointing "up" (as you view it), the other a vertex pointing "down". We call these "u" and "d" holes, respectively. A hole of one type has vertices in common with 3 holes of the second type.

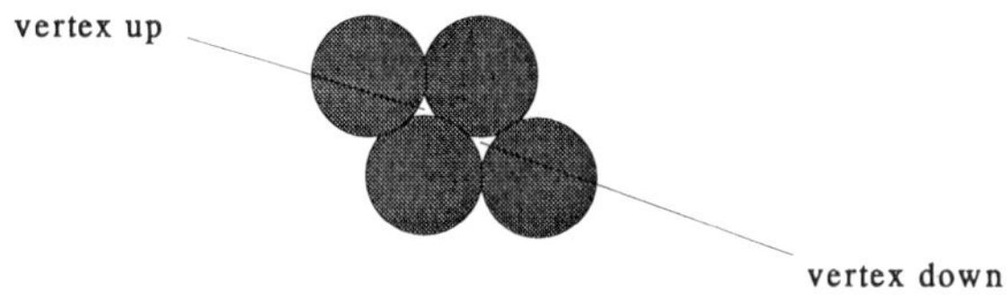

d. Place a blue sphere in the first "d" hole in the upper left corner of the red layer. This begins the closest packing arrangement of spheres in two layers.

e. The blue sphere and the three red spheres that it touches should look as follows:

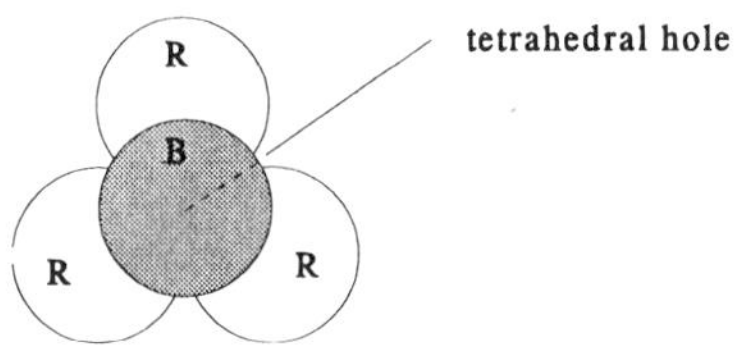

These 4 spheres form a regular tetrahedron (polyhedron with 4 vertices and 4 faces, each an equilateral triangle). The space at the center of the 4 spheres is a <u>tetrahedral hole</u>.

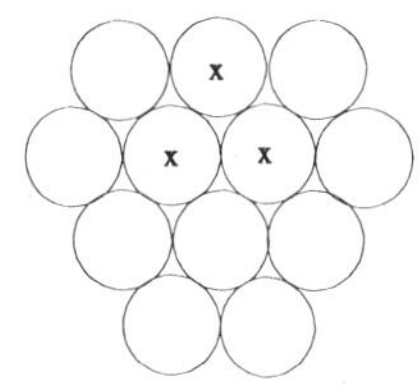

f. Add blue spheres to the second layer, using "d" holes, to give a pattern that looks as shown at right (ignore x's). Construct a second identical blue pattern, separated from the first, in the lower right corner of the red layer. Use only "d" holes. When done, observe that none of the "u" holes in the red layer are used. If you look through the "d" holes in the blue layer, you can see the "u" holes in the red layer.

g. Focus on the equilateral triangle of blue spheres marked by x's in the diagram at right. Immediately under this triangle, find a similar triangle of red spheres rotated 60° with respect to the blue triangle.

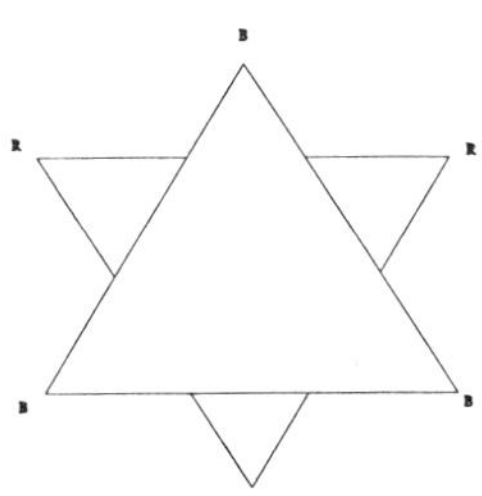

These 6 spheres form a regular <u>octahedron</u> (polyhedron with 6 vertices and 8 equilateral triangle faces), which looks as shown at right. The space at the center of the 6 spheres is an <u>octahedral hole</u>.

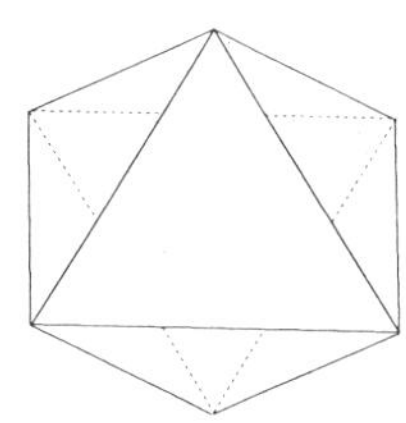

h. There are two ways to place a third layer of spheres above the first two. Using "u" holes in the blue layer will place atoms in rows directly above the red spheres in the first layer. Using "d" holes in the blue layer, neither blue nor red spheres will lie directly below the spheres in the third layer.

1. Using red spheres and the upper left blue layer, place the third layer over the first layer. This gives a central sphere surrounded by a hexagon lying directly over an identical hexagon in the first layer (find it). The two hexagons are separated by an equilateral triangle of blue spheres in the second layer: We characterize this packing arrangement as RBRB..., the letters indicating that the colors and locations of spheres repeat every other layer. This packing arrangement is <u>hexagonal closest packed</u> (hcp). Its more general designation is 1212. . . . Hexagonal channels run vertically through the structure.

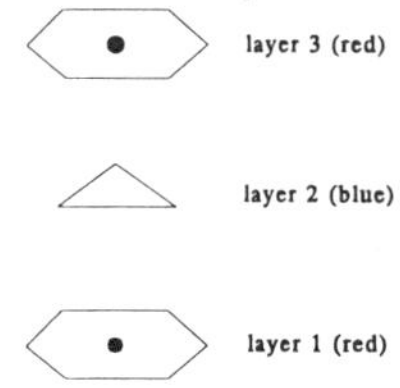

2. Using yellow spheres and the other blue layer, place the third layer so that neither red nor blue spheres lie below the yellow spheres. Use "d" holes of the blue layer. This gives a yellow triangle of 6 spheres, vertex down. Finally, put a red sphere in the center hole of the yellow layer, lying directly over a red sphere in the first layer: This packing arrangement, called <u>cubic closest</u>

packing (ccp), is characterized as RBYRBY. . . or 123123. . . .

Question

1. What is the coordination number (the number of atoms touching an atom) of an atom in a ccp structure? In a hcp structure?

Save these two models for later reference and proceed to the next part.

2. Face-centered cubic lattice. (Steps 2a-2c show that the face centered cubic lattice and cubic closest packing are identical. Be sure you understand everything in a step before proceeding. If you have trouble, get help!)

   a. Use red spheres and the right lower front octant of the kit to construct a unit cell of a face-centered cubic (fcc) lattice (one atom at each corner, one atom at the center of each face of a cube). Do not upset the close packed balls on top! Look along a body diagonal of the cube to see an atom close to you, 6 in a triangular 2nd layer, 6 in a triangular 3rd layer, and one in a 4th layer.

   Look at the cubic closest-packed structure. Looking down on the top red sphere, you will see the yellow triangular 2nd layer, below it a blue triangular 3rd layer, and below that, a red sphere. This shows that the arrangement of atoms in a fcc structure is identical to that in a ccp structure. Do not proceed until you see this.

Question

2. How many nearest neighbors does an atom have in the fcc model? (This should be the same as the coordination number in the ccp structure in section 1.)

---

   b. Add one-half unit cell to the left of the fcc unit cell (think before you build) and remove the right face of your original unit cell. This gives a "new" unit cell with one atom at the center of each edge and one in the center. This is still a fcc structure, viewed differently. Sight down a body diagonal. You will see a triangle of atoms close to you, a second hexagonal layer of 7 atoms, and a third triangular layer turned 60° with respect to the first layer. Do not proceed until you can find this arrangement of layers in your ccp structure.

Question

3. Draw a diagram of the cube with atoms at the edge centers and body center of the cube. Connect atoms

to show the two triangles and hexagon.

Restore the fcc unit cell to standard form (as in part 2a).

c. FIND THE FACE-CENTERED CUBE IN YOUR CCP STRUCTURE! (This is the most important experiment. Do not proceed until you find the cube. If you want help, ask.)

3. Lattice Holes.
   a. You will now find octahedral and tetrahedral holes in the fcc lattice.
   b. Find the hole (space) at the center of the fcc unit cell. This vacancy is surrounded by 6 spheres, equidistant from it and forming a regular octahedron. The vacancy is an <u>octahedral hole</u>!

Question

4. How many octahedral holes are in the unit cell (vacancies at the centers of cube edges are octahedral holes belonging partially to the unit cell).

---

c. Locate the following atoms:

Atom 1 - upper left corner, front face (001)
2 - center, front face (1/2 0 1/2)
3 - center, top face (1/2 1/2 1)
4 - center, left face (0 1/2 1/2)

These atoms form a <u>tetrahedron</u>. The space at the center of this 4-atom group is a <u>tetrahedral hole</u>.

Questions

5. How many tetrahedral holes are there in the unit cell?
6. How many atoms are there in the unit cell? (Some atoms belong only partially to this unit cell!).
7. Do your findings support the statement: there are 2 tetrahedral holes and 1 octahedral hole <u>per atom</u> in a fcc unit cell?
8. Which is larger, a tetrahedral or an octahedral hole?

---

4. Structures of compounds based on the fcc lattice.
   a. Dismantle all models and remove the grooved shelf from the kit.
   b. Use the full dimensions of the kit to construct a face-centered cubic lattice of red spheres.
      1. Rock-salt (NaCl) structure.
         a. Put a blue sphere in each octahedral hole of the unit cell (see section 3b). If the red spheres represent Cl^- ions and the blue Na^+ ions, this is a unit cell of NaCl. The red spheres form a "standard" fcc lattice, the blue spheres a "nonstandard" fcc lattice (section 2b). NaCl consists of two interpenetrating fcc lattices, one of Na^+, the other of Cl^- ions. Na^+ ions go in the octahedral holes of the Cl^- lattice, and vice versa. Many other compounds crystallize with the NaCl structure: alkali halides (e.g., KBr), hydrides (e.g., NaH), some alkaline earth oxides (e.g., CaO), and NH_4Cl (NH_4^+ equivalent to Na^+).

Questions

9. How many Na^+ ions are there per unit cell of the NaCl structure?
10. How many Cl^- ions are there per unit cell?
11. Is this consistent with the 1:1 stoichiometry implied by the formula?
12. What is the coordination number of each cation in the structure?
13. What is the coordination number of each anion?

---

2. Zinc-Blende (ZnS) Structure.
   a. Remove the Na^+ ions (blue spheres) from your structure.
   b. Fill 1/2 the tetrahedral holes (see section 3c) in the red fcc lattice with yellow spheres, so that the yellow spheres form a tetrahedron. If the yellow spheres represent Zn^{2+} ions and the red spheres S^{2-} ions, this represents a unit cell of ZnS. BeO also has this structure.

Questions

14. Confirm the 1:1 ion ratio in ZnS.
15. What is the coordination number of Zn^{2+} (yellow spheres)?
16. What is the coordination number of S^{2-} (red spheres)?

---

3. Fluorite (CaF_2) Structure.
   a. Add yellow spheres to the previous model to fill all tetrahedral holes. With red spheres = Ca^{2+} ions and yellow spheres = F^- ions, the model represents a unit cell of CaF_2.

Questions

17. Confirm the 1:2 ion ratio in CaF_2.
18. What is the coordination number of Ca^{2+} (red spheres)?
19. What is the coordination number of F^- (yellow spheres)?

---

4. Anti-Fluorite Structure.
   a. The compound Na_2O has this structure.

Questions

20. From the name of the structure and the chemical formulas CaF_2 and Na_2O, describe cation and anion positions in the anti-fluorite structure.
21. What are the coordination numbers?

---

5. Diamond Lattice
   a. Remove the yellow spheres and fill 1/2 the tetrahedral holes with red spheres, so the added red spheres form a tetrahedron. If the red spheres = carbon atoms, this represents the unit cell

of <u>diamond</u>. (In diamond, the carbon atoms are bonded covalently, <u>not</u> ionically).

Question

22. What is the coordination number of a C atom in the diamond lattice (check several atoms in your model!)?

---

5. The Body Centered Cubic Lattice.

   a. Using the 3 bottom shelves of the kit, make a body-centered cubic (bcc) unit cell using red spheres (a bcc lattice has one sphere at each cube vertex and one at the cube center). Many elements crystallize with the hcp, ccp, or bcc arrangement of atoms. For some elements, one structure may be stable under one set of conditions and another under a different set of conditions; e.g., iron exists as bcc or ccp, depending on previous heat treatment.

   He, Be, Mg, Zn, and Cd are hcp; the rare gases (except He), Ca, Sr, Ni, and Au are ccp; and the alkali metals, Ba, and the vanadium and chromium groups are bcc.

Questions

23. How many atoms are there per unit cell in the bcc lattice?
24. What is the coordination number of an atom?
25. Can you find any octahedral or tetrahedral holes by extending the lattice?

---

6. Simple Cubic Lattice

   a. Remove the central red sphere from your unit cell. You are left with a <u>simple cubic unit cell</u> (spheres at the corners of a cube), the simplest cubic unit cell.

   b. Put a blue sphere at the center of the cube. With red spheres = Cl^- ions and blue sphere = Cs^+, this represents the unit cell of CsCl. By extending the lattice, convince yourself that CsCl may be viewed as either Cs^+ ions in cubic holes of a simple cubic lattice of Cl^- ions, or Cl^- ions in cubic holes of a simple cubic lattice of Cs^+.

Questions

26. What is the coordination number of Cs^+? Of Cl^-?
27. Calculate the number of Cs^+ and Cl^- ions per unit cell and verify the stoichiometry of CsCl.

---

7. Back to Lattice holes

   a. Use the three bottom shelves to make a fcc lattice at the right side of the kit.

   b. The fcc and ccp arrangements are equivalent. The unit cell just constructed is not actually close-packed, however. It is expanded for better visibility. Imagine that the spheres uniformly move together until spheres along the <u>face diagonal</u> of the cube are just touching, as shown below:

The corner spheres do not touch, but each corner sphere touches the face center sphere. In terms of the sphere radius, r, the face diagonal has length 4r.

c. Find a tetrahedral hole and an octahedral hole. Which is larger?

Whether a cation occupies a tetrahedral or an octahedral hole in a close-packed anion lattice depends on its size compared to the sizes of the holes. In ZnS, Zn^{2+} is small enough relative to S^{2-} to fit tetrahedral holes. In NaCl, Na^{+} is too big for tetrahedral holes, but fits octahedral holes. In CsCl, Cs^{+} is so big compared to Cl^{-} that it cannot fit into either type of hole. The simple cubic anion lattice is adopted.

Questions

28. Based on the considerations in b above, and assuming an anion radius r_o, calculate, in terms of r_o, the radius of a cation that will just fit an octahedral hole in a fcc anion lattice. (Hint: Consider a face of the face-centered cube, and use the Pythagorean Theorem.)
29. Calculate the radius of the cation that can just fit the hole in a simple cubic lattice of anions of radius r_o. (Hint: Consider a right triangle made from a body diagonal, a face diagonal, and an edge of the simple cube.)
30. Calculate the just-fit radius for a tetrahedral hole in a fcc anion lattice. (Hint: Focus on one of the 8 small cubes that make up the fcc unit cell).

When finished, return spheres to boxes.

Disposal Methods

No disposal required.

Preparation
Solids

Name ______________

Section ______________

Date ______________

Read

1. The appropriate sections of your textbook.

Problems

1. Give 4 examples of crystals which you encounter on (almost) a daily basis.

2. Consider the parallelogram defined by the centers of 3 rows of close-packed spheres, 5 spheres per row, shown below:

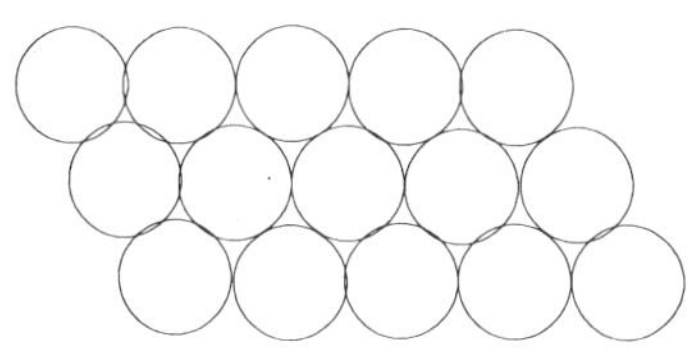

a. Calculate the parallelogram area in terms of sphere radius.

b. Show that the area is smaller than that covered by the centers of spheres arranged as follows:

Results
Solids

Name ______________________
Coworker ______________________
Section ______________________
Date ______________________

Answers to Questions

VSEPR

1 lab period, work in groups. Complete the Preparation page before laboratory.

Goals

- To learn to apply VSEPR theory to real molecules
- To learn to use molecular models

Background

Between 1912 and 1915, William and Lawrence Bragg developed the technique of X-ray diffraction and determined the structures of NaCl and KCl. This achievement earned them the 1915 Nobel Prize in Physics. Since then, X-ray diffraction has been used to study larger and more complex molecules. In 1960, John Kendrew and Max Perutz solved the crystal structure and molecular structure of myoglobin and hemoglobin for which they won the 1963 Nobel Prize. Elucidation of such structures often required months or years of research. For example, Perutz worked for 23 years on the structure of hemoglobin. Today, computers and automated equipment have greatly reduced the time and effort required to determine structures. Many of the structures in your text, together with bond angles and bond distances, are based on X-ray structures. However, if a single crystal cannot be grown, X-ray diffraction is much less useful. Techniques for experimentally determining structures are summarized below.

Technique	Molecular Size Range And Limitations
X-ray diffraction	Small to very large molecules; single crystals only
Electron diffraction	Small to medium molecules
Microwave spectroscopy	Small molecules; gas phase
Neutron diffraction	Small to large molecules; crystals
X-ray absorption	Small to large molecules but only one atom of the absorbing species present per molecule; only a few labs capable.

From accumulated data, generalizations about bond lengths and bond angles have been made. From this information, molecular models of macroscopic size were developed to assist in the visualization of molecules. With the aid of such models and the X-ray crystallographic data obtained by Rosalind Franklin, James Watson and Francis Crick determined the structure of DNA in 1953. This was another Nobel Prize winning effort (1962).

Models still play an important role in research. For example, a chemist may build a model of a molecule whose structure is not experimentally known as an aid to designing a synthesis or to making an informed guess about its structural properties, such as interatomic distances.

You will work with molecular model kits in this experiment. These models help you visualize the arrangement of atoms in space although they will not provide much information about the sizes of the atoms.

Does VSEPR hold for all molecules?
Under what conditions does VSEPR fail?

Equipment and Materials

Molecular model kit

Safety

Eye protection should always be worn in the laboratory.

Experimental

Record your results in your notebook or on the **Results** sheet, whichever is applicable. For each molecule draw the Lewis electron dot structure. Note the positions of any lone pairs. Name the geometry predicted. Then, using the molecular model kit, build a model of the molecule and answer the questions about each molecule or group of molecules.

1. Build models of SF_6, IF_5, and PF_5. Are all of the X-F bonds and F-X-F bond angles in a given molecule identical? Why or why not? If different, how many different kinds are there? Indicate any differences on a sketch.
2. A chemistry major groups BrF_3, XeF_2, SF_4, and PF_5, together. Do the structures have anything in common? Are the X-F bonds and F-X-F bond angles in a given molecule identical? As above, indicate diffences on a sketch of the molecules.
3. Compare the structures of SF_4 and XeF_4. Are they similar? Why or why not? Designate equivalent sets of bonds in each of the structures.
4. CH_4, NH_3, and H_2O are isoelectronic. Demonstrate that this is true. Predict the H-X-H bond angle for each molecule. Justify your answer.
5. Do the structures above agree with those predicted by VSEPR theory? Why or why not?
6. Draw a Lewis dot structure and build a model for benzene, C_6H_6. What would you predict in terms of bond lengths, bond strengths and bond angles based on your model? Typical C-C bond lengths are: single 1.54 Å, double 1.33 Å. In benzene the bond angles and bond strengths are equivalent and the bond length is 1.39 Å.
7. Ozone, O_3, has equivalent O-O bonds. Draw the Lewis structure and build a model for ozone. Compare this to the model of benzene.

Disposal Methods

Return the models to your instructor.

Questions

1. Did VSEPR theory hold for all of the molecules studied?
2. Is there a more appropriate model for these molecules?

References

Adapted from the "Seeing is Believing" laboratory experiment at Amherst College.

Preparation
VSEPR

Name ______________
Section ______________
Date ______________

Read

1. This experiment.
2. VSEPR in your text.

Questions

Define the following terms:

VSEPR
lone pair electrons
bonding electrons
isoelectronic

Results
VSEPR

Name __________________

Coworker(s __________________

____

Section __________________

Date __________________

Lewis Dot Structures

Answers to Questions

Symmetry

1 lab period; work in pairs. Complete the Preparation page before laboratory.

Goals

- To gain experience in building models of covalent species
- To practice making visually meaningful drawings of molecules
- To learn to recognize the symmetry elements possessed by a molecule

Background

Most of us have a qualitative understanding of symmetry, and we use this concept in making judgments in everyday life. For example, we look for symmetry in Christmas trees, gemstones, and automobiles. When scientists speak of symmetry, however, they have much more than qualitative considerations in mind. Consider 3 objects: a sphere, a cube, and a planar hexagon. Which of these objects is most symmetrical, and which least? Clearly the sphere has the most symmetry and the hexagon the least, but why? The answer is simple. There are many ways that we can orient a sphere so that it looks the same. We call these equivalent orientations, because they cannot be distinguished from one another. There are an infinite number of equivalent orientations for a sphere. For a cube, there is a large but finite number of equivalent orientations, but not all orientations are equivalent. Consider the two in Figure 1. We can easily distinguish these, so the orientations are not equivalent. Finally, the planar hexagon has the least number of equivalent

Figure 1

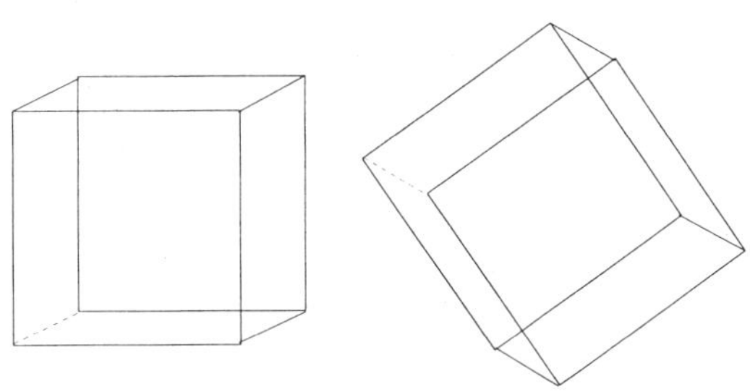

orientations. We "measure" the symmetry of an object according to the number of equivalent orientations which we can find for it. Unfortunately, it is not particularly useful to know the number of equivalent orientations for an object. It is much more useful to consider the types of movements which are used to convert one orientation to an equivalent one.

There are several ways to move an object from one orientation to another:

> rotation about an imaginary axis (a straight line) which passes through its center of gravity;
> reflection in an imaginary plane which passes through its center of gravity;
> inversion through its center of gravity (an imaginary point).

If one or any combination of these movements produces an orientation of an object indistinguishable from (appearing the same as) the original orientation, the movement is called a symmetry operation (abbreviated SO). The geometrical entity (point, line, or plane) about which the movement is carried out is called a symmetry element. Three symmetry elements and the corresponding SOs are presented in Table 1.

Table 1
Three Common Symmetry Elements

Symmetry element	Symbol	SO
plane of symmetry	σ	reflection in the plane
proper rotation axis (a line)	C_n	rotation about the axis by the angle 360/n or an integral multiple thereof
center of inversion (a point; also called a center of symmetry)	i	inversion through the center

We now consider examples of each symmetry element and its corresponding SOs, using simple 2- or 3-dimensional figures for illustration.

The plane of symmetry

Figure 2

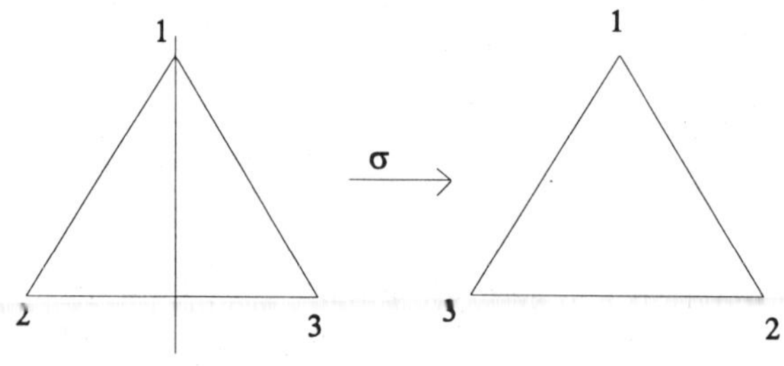

A plane of symmetry is a plane which divides an object into two identical halves. The equilateral triangle in Figure 2 has 4 planes of symmetry. The line through the top vertex and bisecting the base indicates the position of one symmetry plane, which contains this line and is perpendicular to the plane of the paper. To show that this is a symmetry plane, we reflect the triangle in it, and show that the resulting orientation of the triangle is indistinguishable from the original. To reflect the triangle, imagine the plane to be a mirror. When the triangle is reflected in the mirror, every point originally to the left of the mirror will be moved to the corresponding position on the right, and vice versa. Instead of considering the effect of reflection on every point in the triangle, we need consider only its effect on the prominent points (i.e., the vertices). Thus the left vertex will move to the right side, the right vertex will move to the left side, and the top vertex, which lies in the mirror, will not be moved. To show the effects of an operation it will be convenient for us to label the prominent points of an object (e.g., the vertices of our triangle) with numbers. In performing the operation, we move the numbers with the points to show the new orientations. We can then readily tell if two orientations are equivalent or identical. Keep in mind that these numerical labels are only bookkeeping devices by which we track the movements of an object. They are not an inherent part of the object. We put numbers on the triangle and do the reflection as illustrated in Figure 2. The movements of the numbers correspond to our verbal description of the reflection operation. Note that but for the numbers, the orientations of the triangle before and after reflection would be indistinguishable. Thus this reflection is a symmetry operation, and the plane in which it is performed is a symmetry element. Generally, we use the symbol σ to represent both a symmetry plane and the SO of reflection in this plane. The σ above the arrow tells us to reflect in the indicated plane.

The remaining 3 symmetry planes of the triangle are easy to find. There are two more just like the first one, passing through vertices 2 and 3. The 4th is the plane of the triangle. Reflection in this plane leaves all vertices unmoved, since all three vertices lie in the plane.

Finally, what happens if we do two successive reflections in a plane (Figure 3)? The final orientation is identical to the original. Performing two successive reflections in the same plane is like doing nothing at all!

Figure 3

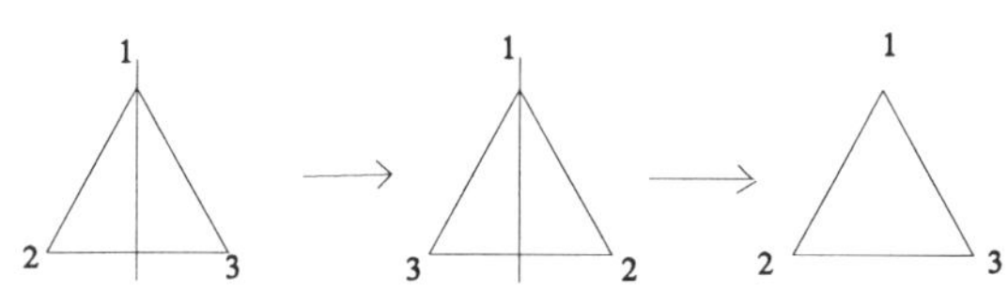

The proper rotation axis, C_n

Consider a line perpendicular to the plane of a square and passing through its center. If we rotate the square by 90° (or integral multiples of 90°) about this line (axis), we produce figures indistinguishable from the original. For consistency, we use clockwise (CW) rotations. Then the rotations of the square are pictured in Figure 4. Four successive CW 90° rotations give an orientation identical to the original. We symbolize

Figure 4

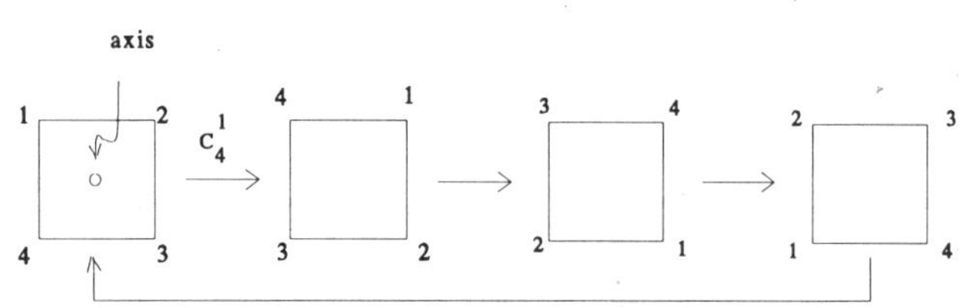

the proper rotation axis C, followed by a numerical subscript, n. n is the order of the axis, and is obtained by dividing 360° by the smallest angle through which we must rotate to produce an equivalent orientation. In this case, the angle is 90°, so the axis is C_4. The square has a fourfold axis of proper rotation, or a proper rotation axis of order four. As for reflection, the symbol C_n represents both the symmetry axis and the operation of clockwise rotation by the angle 360/n. We indicate rotation of the square clockwise by 90° about its fourfold axis as shown in Figure 4.

Two successive rotations by 90° each is equivalent to a single rotation of 180°. To indicate two successive 90° rotations about the C_4 axis, we write

$$C_4^2$$

where the 2 tells us to perform the C_4 operation twice. A 180° rotation may be symbolized C_2, so

$$C_2 = C_4^2$$

3 successive 90° rotations about the C_4 axis is equivalent to a single rotation by 270°, and is written C_4^3; 4 successive 90° rotations is symbolized C_4^4. This operation, rotation by 360°, is equivalent to doing nothing. Thus we see that a single C_4 axis (the symmetry element) generates the following SOs:

$$C_4 \qquad C_4^2\,(= C_2) \qquad C_4^3 \qquad C_4^4$$

In addition to the C_4 axis, the square has four C_2 axes perpendicular to the C_4. Two of these are indicated in Figure 5 (note that we use a line to represent either an axis or a plane of symmetry). Can you locate the other two?

Figure 5

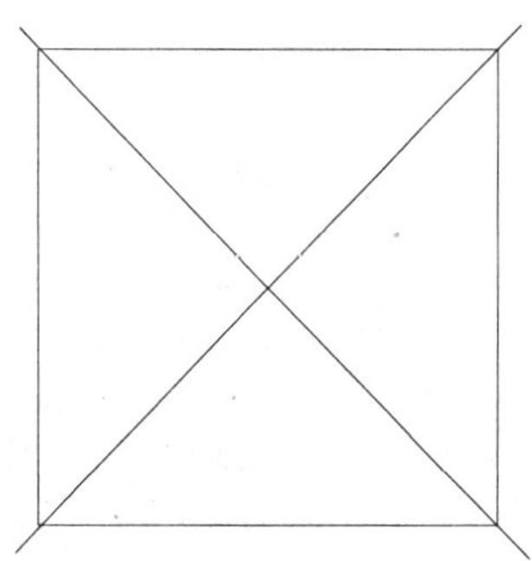

The center of inversion (center of symmetry), i

Imagine the center of the square to be the origin of a Cartesian coordinate system, and consider an operation in which we change the coordinates (x,y,z) of every point in the square to (-x,-y,-z). This is equivalent to moving every point on the square through the center of the square an equal distance to the opposite side. This operation, called inversion, produces an orientation equivalent to the original, and is thus a symmetry operation for the square. We symbolize both the inversion center and the SO of inversion with the symbol i, as in Figure 6. Two successive inversions of an object give an orientation identical with the original (try it with the square).

Figure 6

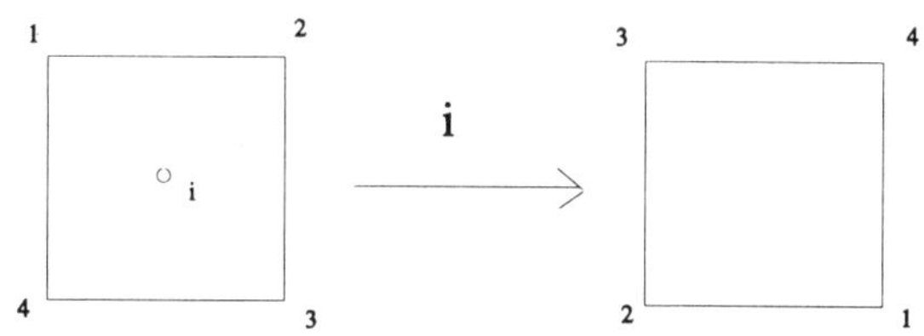

Not all regular geometric figures have a center of inversion. Consider inverting an equilateral triangle, as in Figure 7. We can distinguish the final orientation from the original, so inversion is not a symmetry operation for the equilateral triangle.

Figure 7

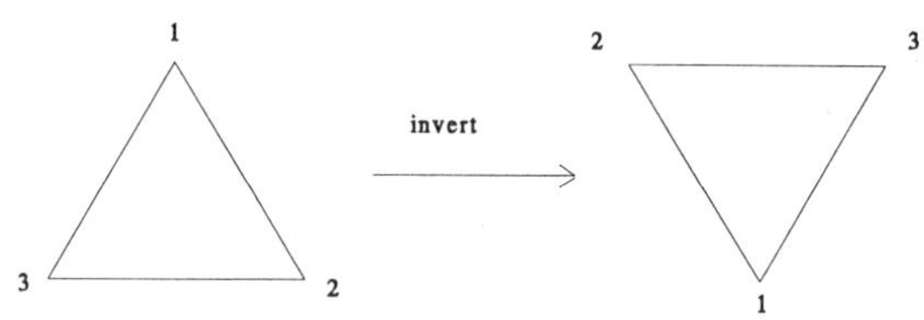

The symmetry of molecules

Since the shapes of molecules are usually described in terms of geometrical shapes, no new considerations are needed to perform SOs on molecules. For example, BF_3, which is trigonal planar, has the same symmetry properties as an equilateral triangle; and the ion ICl_4^- has the same symmetry properties as a square (Figure 8). All of the symmetry elements of the triangle are listed below.

Proper axes: C_3 perpendicular to plane of triangle

$3C_2$'s perpendicular to the C_3 and lying in the plane of the triangle. Each passes through one vertex and the center of the triangle

(We are listing symmetry elements, not SOs, here, so C_3^2, a SO generated by the C_3 axis, is not listed.)

Symmetry planes: 3 planes perpendicular to the plane of the triangle, each containing a vertex and bisecting the opposite side.

The plane of the triangle.

Center of symmetry: no.

We now bring in a few more conventions. The axis of highest order is called the <u>principal rotation axis</u>. For the triangle, the principal axis is C_3. Symmetry planes which contain the principal axis (i.e., having the principal axis lying in them) are called <u>vertical planes</u>, designated σ_v. Symmetry planes perpendicular to the principal axis are called <u>horizontal planes</u>, designated σ_h. In shorthand, the symmetry elements possessed by the equilateral triangle are:

$$C_3 \qquad 3C_2 \qquad 3\sigma_v \qquad \sigma_h$$

Figure 8

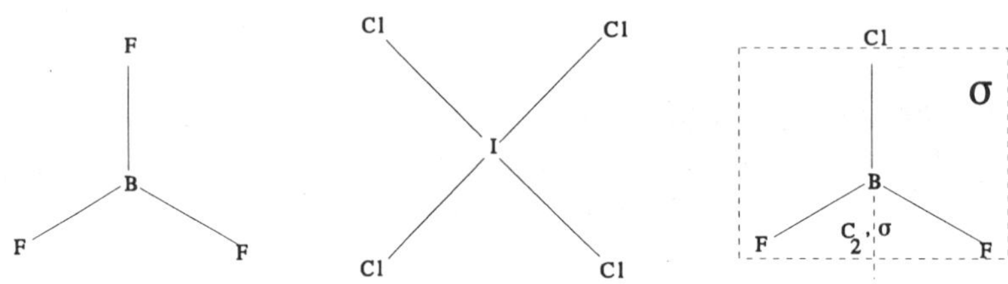

The equilateral triangle, BF_3, and any other molecule or object possessing this set of symmetry elements have the same symmetry, and are grouped together. The molecules BCl_3 and BBr_3 are in this symmetry group.

Suppose we replace one of the F atoms in BF_3 with chlorine to obtain BF_2Cl. Does this belong to the group of the equilateral triangle? No. This new molecule has less symmetry than BF_3. In fact it has only one C_2 axis and two vertical symmetry planes, indicated in Figure 8. One of the symmetry planes contains this axis and is perpendicular to the plane of the molecule. The other is the plane of the molecule.

Compiling a complete list of the symmetry elements for a molecule seems an imposing task, especially to one who is just learning to do it. How do you know when the list is complete? Although we cannot give a foolproof way of finding them all, a few guidelines based on geometrical principles can be stated.

1. If you find a C_n axis, $n \geq 2$, and one C_2 axis perpendicular to it, there are n-1 more such C_2 axes. For example, the square has a C_4 axis and 4 C_2 axes perpendicular to it.

2. If you find C_n and one vertical plane of symmetry, there are n-1 more σ_v's; i.e., a molecule with C_n must contain either zero or n σ_v's.

3. If you find C_n, n even, and σ_h, there must be a center of symmetry.

4. Along any line where two planes of symmetry intersect, there is an axis of symmetry. For example, BF_2Cl has two planes of symmetry (the plane of the molecule and a plane perpendicular to the plane of the molecule along the B-Cl bond) which intersect along the B-Cl bond. This line must be an axis of symmetry, and inspection shows it to be C_2.

5. If two axes of symmetry intersect, there must be a plane of symmetry containing both axes.

6. Some characteristics of molecules containing symmetry planes:
   a. Atoms which occur only once in the structure must lie in all σ;
   b. If an atom occurs twice in the structure, any σ must either <u>contain</u> or <u>perpendicularly bisect</u> the line joining the two like groups;
   c. All atoms of a given type which do not lie in the plane must occur in even numbers.

7. Some characteristics of molecules having a C_n axis:
   a. There may be any number of species of atom lying on an axis, because rotation does not shift them.
   b. If one atom of a certain type lies off C_n, there must be n-1 more such atoms at equal intervals about the axis, for a total of n.

8. Some characteristics of molecules having a center of symmetry:
   a. No molecule can have a center of symmetry unless the polyhedron whose geometry it adopts has a center of symmetry.
   b. An atom which occurs only once in a molecule must be at the center of symmetry. All other atoms occur in pairs.
   c. A molecule with an odd number of more than one atom cannot possess a center of symmetry.

In this experiment you will build models of several molecules, diagram each molecule on paper, and make a list of its symmetry elements. Finally, based on the lists of symmetry elements, you will classify the molecules into a small number of groups. Several guidelines will help you in making your drawings. Usually a molecule is drawn with as many atoms as possible lying in the plane of the paper. Bonds between these atoms are represented with solid lines. Bonds to atoms lying <u>behind</u> the plane of the paper are represented using dashed lines, and bonds to atoms <u>above</u> the plane of the paper are represented by wedges, intended to indicate a bond projecting toward the viewer. The structures of tetrahedral methane, trigonal bipyramidal PF_5, and octahedral SF_6 are shown in Figure 9.

Figure 9

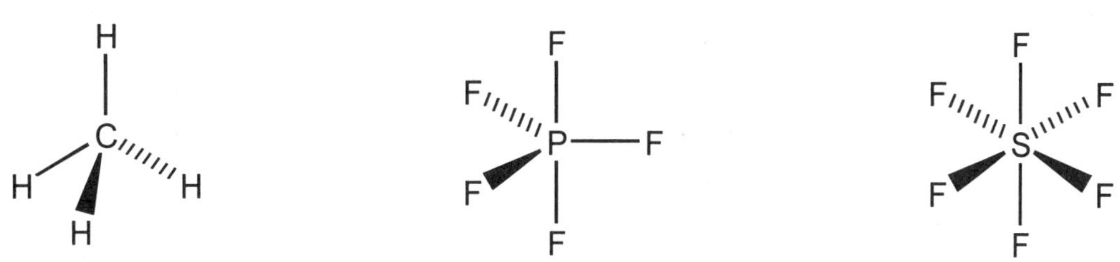

Equipment and Materials

1 molecular model kit.

Experimental

Record all observations in your lab notebook or on the **Results** sheet, whichever is applicable. Obtain a molecular model kit. We will consider simple molecules, with a central atom and between 2 and 6 <u>terminal</u> atoms, or ligands. Your instructor will describe the components of the molecular model kit. Begin by sorting your kit into groups of like pieces.

Molecules having three electron domains (i.e., electron groups) around the central atom.

Molecules in this category are

BF_3 (B is the central atom)
$:SnCl_2$ (Sn is the central atom)

For each, use VSEPR theory to predict (1) the electron group distribution about the central atom and (2) the shape of the molecule (based on atom positions only). Build a model of each. Draw each molecule, then find as many symmetry elements as you can for each (look for planes first, then axes, then a center). Record these using shorthand notation. According to VSEPR, these two molecules have the same electron pair distribution. Do they have the same symmetry elements? KEEP THE MODELS.

Four electron domains

Many molecules fall into this category. Lewis structures of a few examples are given in Figure 10. (The Lewis structure is not intended to convey the shape of the molecule).

Figure 10

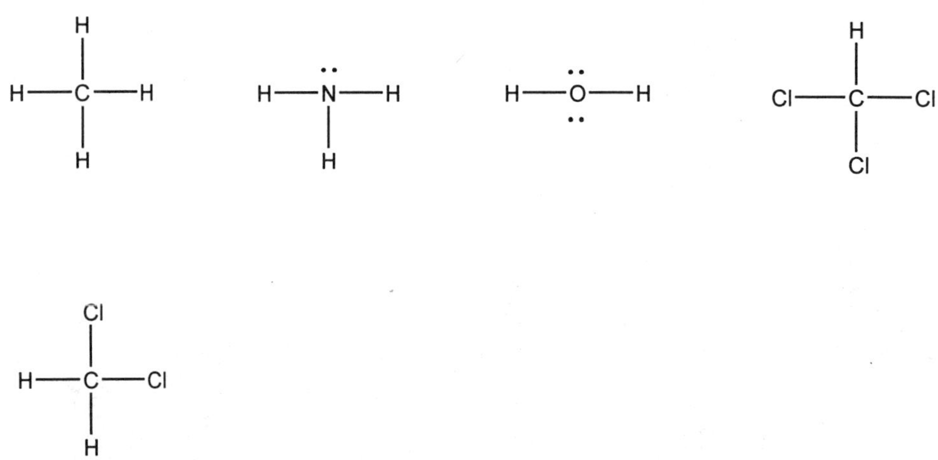

For each molecule, predict the electron group distribution about the central atom and the shape using VSEPR, and record predictions. Build models of CH_4, NH_3, and H_2O. Draw each molecule and list its symmetry elements (CH_4 has a total of 7 rotation axes and 6 symmetry planes. Does it have a center of symmetry?)

When done with this series, build CH_3Cl by replacing one of the four H's in methane with a Cl (use a different colored terminal atom). List the symmetry elements of CH_3Cl. Does CH_3Cl look like NH_3? Do these two molecules belong to the same symmetry group?

Finally, convert CH_3Cl to CH_2Cl_2 by altering a second terminal atom. List the symmetry elements of CH_2Cl_2 in shorthand notation. Does CH_2Cl_2 bear any resemblance to H_2O? to $SnCl_2$? KEEP THE

MODELS OF CH_2Cl_2, NH_3, and H_2O.

Five electron domains

Again, many molecules fall into this category. A few examples are given in Figure 11. Build models of PF_5 and SF_4, based on VSEPR considerations. Two models for SF_4 are conceivable, one with the lone pair equatorial, the other with it axial. Spectroscopic experiments reveal that SF_4 has a C_2 axis and two symmetry planes. Does this information help you decide which structure is correct? Dismantle the incorrect model of SF_4. Which previous model does each of your remaining structures resemble? Record the symmetry elements for PF_5 and SF_4.

Figure 11

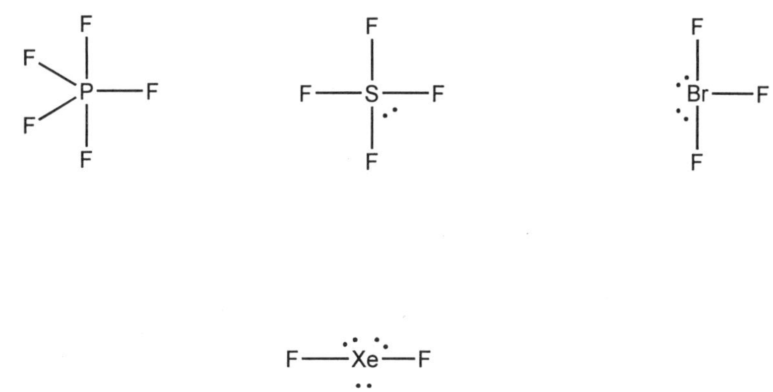

Build PF_4Cl from PF_5 (dismantle one of your other models if you need to). With which other molecules would you classify this molecule?

Build a model of BrF_3. Now there are two lone pairs. How many distinct structures can you build. Decide which is correct, using the fact that spectroscopy tells us that BrF_3 has one C_2 axis and two symmetry planes. Based on the symmetry elements in BrF_3, classify it with other molecules previously built.

Six electron domains

A few examples of species in this category are:

SF_6
XeO_2F_4
BrF_5
ICl_4^-
XeF_3^- (as yet hypothetical)

Predict the electron group distribution and the shape of each, and record your predictions. Build models of SF_6 and BrF_5 and draw them. List as many symmetry elements as possible for each (SF_6, which is perfectly octahedral, has 3 four-fold axes, 4 three-fold axes, 6 two-fold axes, 9 symmetry planes, and a center of symmetry). Orient your model to get each perspective in Figure 12. This should help you find some of these elements.

Build XeO_2F_4 from the SF_6 model and ICl_4^- from BrF_5 How many ways are there to locate the 2 lone pairs in ICl_4^-? This molecule is found by spectroscopy to have a C_4 axis, 4 C_2 axes, 5 symmetry planes, and a center of symmetry. Which structure is correct? Note similarities between XeO_2F_4 and ICl_4^-. Do they have the same symmetry?

Build XeF_3^- from the model of ICl_4^-. Classify this moleçule with other molecules previously examined.

Figure 12

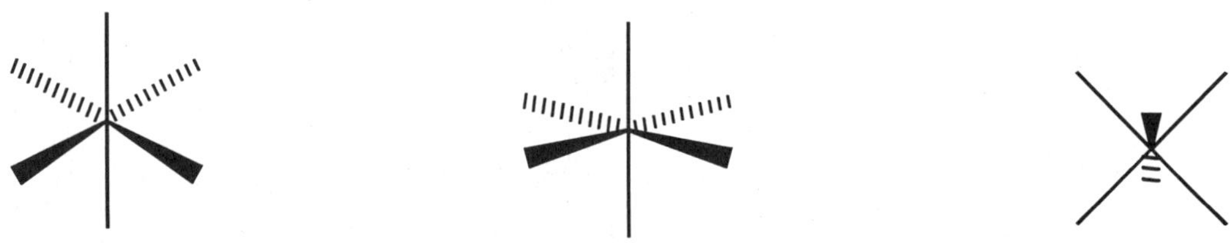

Last, return to XeO_2F_4. Although this molecule exists only with the <u>trans</u>-structure, the <u>cis</u>-structure might be considered possible. Build the cis-structure and classify it with other molecules previously built.

Two electron domains

Examples of species with only two groups of electrons are shown in Figure 13. VSEPR predicts that such species should be linear. Build $BeCl_2$ and SCN^- and find as many symmetry elements as you can for each. Focus on the axis collinear with the molecular axis. Is this an axis of symmetry? What is its order? Are there any C_2 axes in either of these molecules? How many? How about vertical or horizontal symmetry planes?

Figure 13

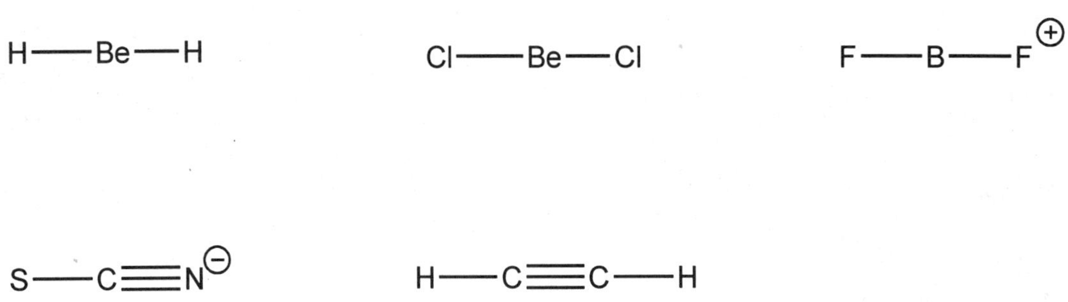

To complete the workshop, classify all of the species which you have built into symmetry groups. All species with the same symmetry elements belong in the same group. For example, both $SnCl_2$ and H_2O have the following symmetry elements:

C_2 $2\sigma_v$

Calling the group of all molecules with these symmetry elements group 1, then $SnCl_2$ and H_2O both belong to group 1. Classify the other molecules considered in this experiment into symmetry groups, which you may label group 1, group 2, etc. and record results. Group 1 has been entered, but be aware that molecules other than $SnCl_2$ and H_2O may belong in this group. How many different groups do you find?

When you have completed the model construction and answered the questions, disassemble all models and return the kit before leaving lab.

Disposal Methods

No disposal is required.

Preparation
Symmetry

Name ____________________
Section ____________________
Date ____________________

Read

1. This experiment
2. The appropriate sections of your textbook

Problems

1. Show the result of each of the following reflection operations.

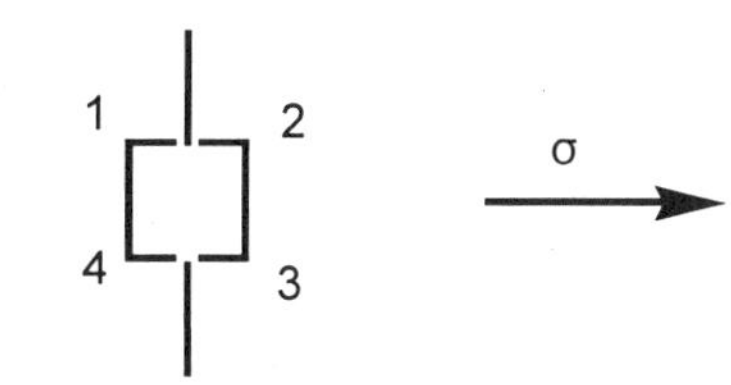

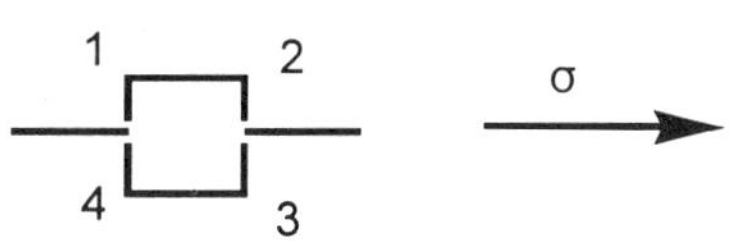

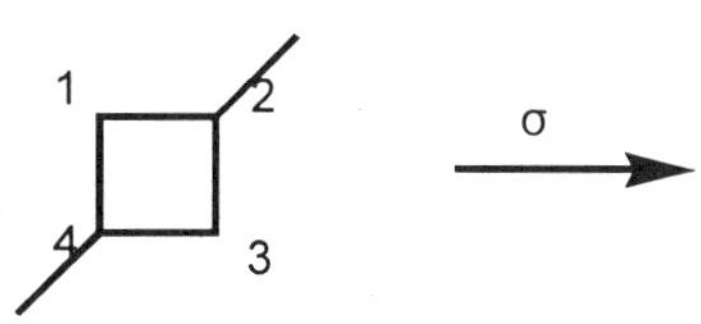

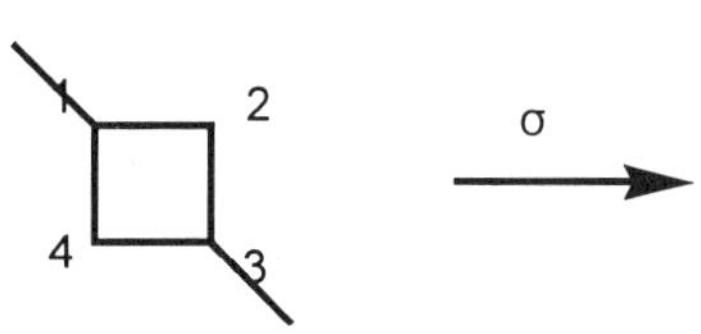

2. What orientation results from each rotation?

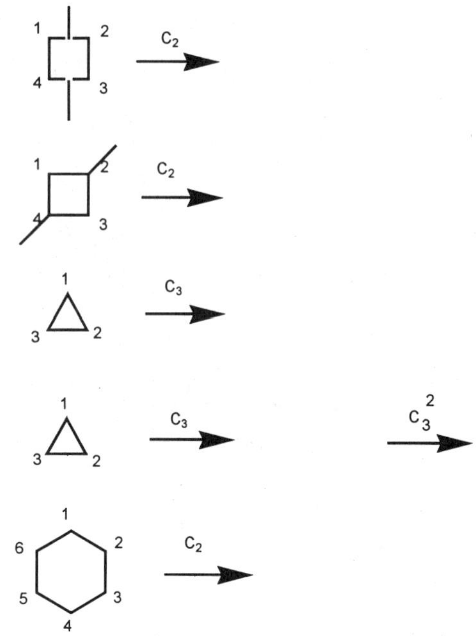

3. Which figures have a center of inversion?
   a. a regular pentagon
   b. a regular hexagon
   c. a cube
   d. a tetrahedron
   e. an octahedron

4. Label the vertices of a regular octagon with numbers and carry out the inversion operation on the figure.

5. List all of the symmetry elements of the square planar molecule XeF_4, using the shorthand notation presented in the **Background** section.

6. Carry out the operation indicated below (the axis passes through B and is perpendicular to the paper). Is C_3 a SO for BF_2Cl? Why or why not?

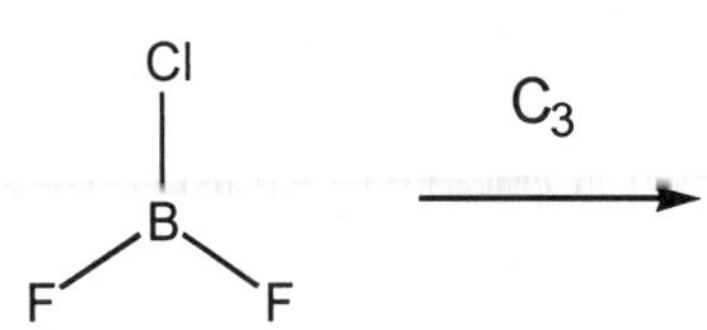

Results
Symmetry

Name ____________________
Section ____________________
Date ____________________

Molecule	e Group Distribution	Shape	Sketch	Symmetry Elements

Molecule	e Group Distribution	Shape	Sketch	Symmetry Elements

Symmetry Group Table

Group	Symmetry Elements	Molecules
Group 1	$C_2, 2\sigma_v$	H_2O, $SnCl_2$

Projects

Beverages†

2 lab periods; (written proposal and brief 1 to 2 page summary of results; oral report;) work in groups of 2 or 3. Complete the Preparation page before laboratory.

Goals

- To design a research project for your beverage(s)
- To learn to write a proposal for a research project

Background

You will select a beverage or beverages to study, e.g., coffee or tea. Define the questions you wish to answer with respect to the beverage. For example, "How much caffeine is in decaffeinated coffee compared to regular coffee?" or "What is the acidity of tea as a function of brewing time and/or brand?"

Once you have defined the question(s) you wish to explore, devise the experiments necessary to answer those questions. For the examples above, you would need to decide what is the best way to measure caffeine in coffee, or what is the best indicator to use when titrating tea.

As you consider your project, include background information to justify the study you are proposing. For instance, what are the health benefits or risks of consuming caffeinated beverages? The background information should be both informative and support your work.

During this course, you have learned various aspects of doing chemistry in the real world. Keep in mind that in this project you will be carrying out experiments on real world samples. The amounts of reactants cannot be known exactly in advance and so part of your planning and experimentation will be determining how much of various reagents and beverages to use. This part of the experimentation in should be considered part of the learning process and used to repeat the experiment in a more productive way.

Equipment and materials vary depending on the experiments selected. Read the experiments and provide your instructor with a list of required materials and equipment.

What do you hope to learn about the beverage(s) studied?

General Considerations

- Solids in samples may block pipet tips. Filter beverages containing solids through cheese cloth before pipetting.
- Colored samples may interfere with observations of reactions and end points. For example, the rose pink endpoint in the determination of vitamin C may be difficult to see in a cranberry juice sample but is observable in orange and grapefruit juice. Most beverages can be decolorized by the addition of activated charcoal. See **Decolorizing Samples**.
- These analyses are quantitative and therefore you will want to use the most accurate and precise measurements available. You should be able to comment on the accuracy and precision of the experiments in your project.

---

†Portions of this project were adapted with permission from Experience the Extraordinary Chemistry of Ordinary Things, Thomas Chasteen and Bruce Richardson, Copyright John Wiley and Sons, Inc., 1992.

•Record all data in your notebook or on the **Results** pages, as instructed.

Suggested Beverages

Juices: orange, grapefruit, cranberry, apple, grape (white and purple)
Sodas: Pepsi®, Mountain Dew®, Crystal Pepsi®, Diet Pepsi®, Jolt®, Coke®, Diet Coke®
Wines: red and white
Beers: regular, lite, nonalcoholic, ale
Ciders: regular, clear, sparkling, hard (may be seasonal)

Determination of Vitamin C[1]

This analysis involves the use of titration. Recall that a titration is an analytical technique used to add a stoichiometric amount of a standard reagent to a second reagent. From the amount of the titrant added, the amount of analyte in the sample is found. Thus, you will be adding standardized 2,6-dichloroindophenol to the beverage to find the amount of vitamin C in the beverage. Note that volumes of titrant added should be between 10 and 40 mL and have 4 significant figures. Appropriately adjust either the titrant concentration or the volume of your sample so that your titration volumes are reasonable.

The following procedure is based on that used by the U.S. Food and Drug Administration. The titrant is the oxidized form of 2,6-dichloroindophenol in a slightly basic solution which has a dark blue color. As the 2,6-dichloroindophenol is added to the vitamin C (ascorbic acid) it is reduced to a colorless compound. Once the ascorbic acid has completely reacted, the excess 2,6-dichloroindophenol, in the oxidized form, reacts with excess acid in the analyte to yield a rose pink color, signalling the end of the titration. The overall reaction between 2,6-dichloroindophenol and ascorbic acid is shown below.

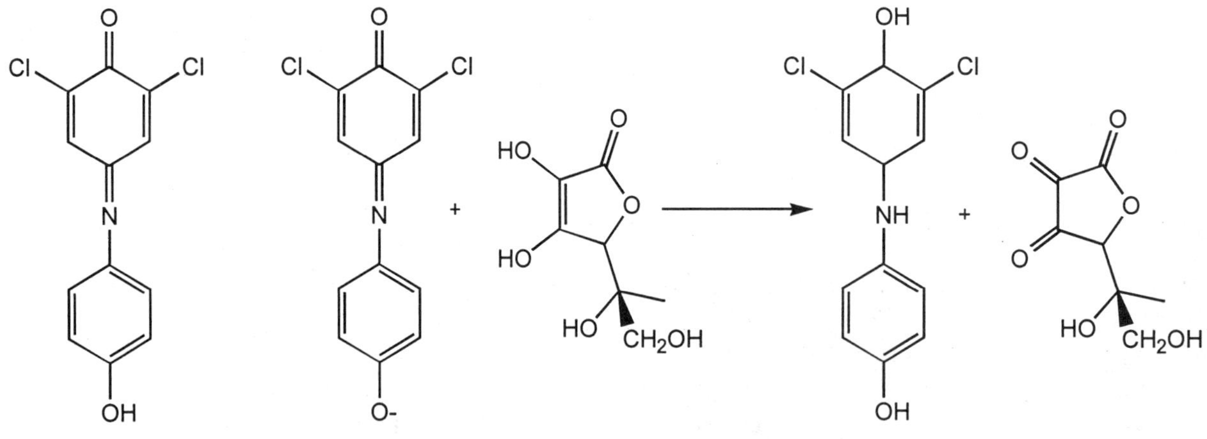

2,6-dichloroindophenol Oxidized form Red in acid | 2,6-dichloroindophenol Oxidized form Blue in base | L-ascorbic acid Vitamin C | 2,6-dichloroindophenol Reduced form Colorless in acid | L-dehydroascorbic acid

Because this procedure requires the sample to be over-titrated by a small volume, the amount that is over-titrated must be corrected. This is done by titrating a blank. The blank is a sample of the background, in this case a solution of metaphosphoric acid/acetic acid reagent. A volume of a 1:1 dilution of the reagent equal to the volume of beverage titrated is titrated. For example, if you find you are titrating 25.00 mL samples of beverage, titrate 25.00 mL of a 1:1 dilution of the metaphosphoric acid/acetic acid background. The volume of titrant needed to titrate the blank is subtracted from the volume of titrant needed to titrate the beverage sample.

Safety

Eye protection must be worn at all times. Acids burn the skin. Wash any spills immediately and inform the instructor. 2,6-Dichloroindophenol stains the skin.

Experimental

Add 50.00 mL of the beverage to a 100-mL volumetric flask and dilute to volume with metaphosphoric acid/acetic acid reagent. Titrate 10.00 to 20.00 mL samples as rapidly as possible with 2,6-dichloroindophenol to a rose-pink endpoint that lasts at least 5 seconds. See note above about titrations.

A blank must be titrated to estimate the amount of 2,6-dichloroindophenol reacting with the acidic background. Dilute the metaphosphoric acid/acetic acid reagent 1:1 with distilled water. Titrate a volume equal to the volume of beverage titrated above. For example, if the volume of beverage titrated is 15.00 mL, titrate 15.00 mL of the 1:1 diluted metaphosphoric acid/acetic acid reagent. This blank is used to correct all of the titration volumes.

The 2,6-dichloroindophenol solution must be standardized the same day as the sample and blank are titrated. 2,6-Dichloroindophenol is standardized with a known sample of ascorbic acid. Dilute approximately 0.07 g (weighed to the nearest 0.1 mg) of ascorbic acid to 250.00 mL with the 1:1 dilution of the metaphosphoric acid/acetic acid reagent. Note that standardized ascorbic acid solution may be provided. Titrate 10.00 to 25.00 mL samples as rapidly as possible. Calculate the concentration of 2,6-dichloroindophenol in mg vitamin C/mL of 2,6-dichloroindophenol. Estimate the precision. Report results in mg vitamin C/mL of beverage or other units which can be readily compared to values reported in the literature.

Disposal Methods

Place all solutions in properly labeled waste beakers or bottles. Do not flush down the drain.

Questions

1. What is the vitamin C content of a "serving" of the beverage?
2. How does this relate to the recommended daily allowance (RDA)?

Determination of pH and Acidity

Acids serve various functions in beverages. Some are naturally occurring while others are added for flavor enhancement or as preservatives. Consider the acid(s) present in the beverage(s). Is there one or more? Do you expect the beverage to be acidic or basic? What is the function of acidity in your beverage(s)? Are the acids in your beverage(s) natural or added?

Some beverages, such as orange juice or grapefruit juice, can be titrated to a phenolphthalein endpoint to determine the total amount of titratable acid present. In other beverages, such as colas, a pH titration is required because the color change at the endpoint is not visible. Finally, some beverages, such as cranberry juice, have internal indicators. These are pH sensitive compounds which will change color as the titration progresses and act as indicators. After considering these points, decide which type of titration is most appropriate for your beverage(s).

In some beverages, the total amount of acid that is titrated will include ascorbic acid and citric acid. If a vitamin C analysis is carried out in conjunction with the acid-base titration described here, the amount of vitamin C is subtracted from the total acid determined to find the amount of citric acid in the beverage. Similar treatments are possible in some beverages which have more than one acid present.

Safety

Eye protection must be worn at all times. Wash any spill of base with soap and water.

Experimental

Conduct an appropriate titration of a sample of your beverage(s). See **Acid-Base Titrations** in this manual for information about setting up a pH titration. Recall that a titration should use 20 to 30 mL of titrant. You may have to quantitatively dilute the base used to titrate the beverage(s). Alternatively, you may have to adjust the volume of beverage(s) titrated.

Disposal Methods

Flush solutions down the drain with water unless otherwise directed by the instructor. Rinse burets and fill with distilled water when done.

Questions

1 Can you distinguish the endpoints for the different acids in your beverage?
2. Calculate a concentration for each acid present. Compare these values with any literature values you can find.

Decolorizing Samples[10]

Colored species can be removed from the sample by the addition of activated charcoal. For solutions which will be diluted before analysis, consider if they will be sufficiently decolorized by dilution. For example, most solutions are decolorized sufficiently by dilution for the sugar analysis. The rose colored endpoint in the vitamin C titration is visible in orange and grapefruit juice without decolorizing first.

Safety

Eye protection must be worn at all times. Before lighting a Bunsen burner, check with your instructor.

Experimental

Consider the total volume of beverage needed before beginning. Decolorize about 10 to 20 % more. Add a small amount (the tip of a small spatula) of the charcoal to the sample, swirl, and filter by gravity through fluted filter paper. If the solution is not sufficiently decolorized, repeat the procedure. For solutions which are not easily decolorized, heat the beverage to boiling, then very carefully add a small amount of the charcoal. Filter the boiling solution.

Disposal Methods

Filter paper may be thrown in the trash unless otherwise directed by the instructor.

Questions

1. How does charcoal act as a decolorizing agent?

Determination of Caffeine by Extraction[5-8]

Note: This is a 2 week experiment.

Caffeine is found naturally in many beverages and is added to many others. Before you begin, consider how much caffeine you expect in a given volume of your beverage. Be sure to start with a large enough volume of beverage that you will obtain a measurable quantity of caffeine. A lower limit for the amount of extracted caffeine might be 50 to 60 mg.

Caffeine is removed from the beverage based on the fact that caffeine is more soluble in dichloromethane (CH_2Cl_2) than it is in water. Extraction occurs when one substance is removed from a solution by its more favorable solubility in a second solvent. Thus, caffeine is being extracted from the aqueous beverage into the dichloromethane in this experiment.

Once the product is isolated, it must be identified as caffeine. The confirmation of the products of a reaction or procedure is good lab practice. Two methods are presented for the confirmation of caffeine. The first is the melting point of the product. A pure substance has a distinct and generally narrow melting point range, i.e., the compound melts over a 1 to 3 degree temperature range. When impurities are present, they lower the melting point and usually broaden the melting temperature range. Both the lowering and the broadening of the melting can be quite extreme depending on the impurity present.

A second method to confirm the presence of caffeine in the product is thin layer chromatography, TLC. The extraction product(s) is dissolved in a small amount of dichloromethane. A sample of pure caffeine is also dissolved in dichloromethane. A small dot of the known caffeine and one of the product are spotted onto the TLC plate. The plate is developed in a solvent mixture. The solvents move up the TLC plate by capillary action and when the solvent front is almost to the top of the plate the development is stopped. Each component in the product will have a different affinity for the plate and thus will move a different distance with the solvent. By comparing the movement of each component with the movement of the known caffeine, caffeine in the product can be identified. A UV light is used to illuminate the plate and mark how far the solvent and each component in the product moved.

To compare the spots on the plate, an R_f value is calculated for each spot. First, measure the distance the solvent front moved by measuring the distance from the original spotting line to the top of the solvent front. This is the $R_{solvent}$ value. Next, measure how far each spot traveled from the spotting line, the R_{sample} value. Then, calculate the R_f value for each spot in the beverage sample and compare them to the known caffeine spots on the same plate. R_f is the ratio $R_{sample} / R_{solvent}$. Any spot with an R_f value equal to the known caffeine R_f value is identified as caffeine.

Safety

Dichloromethane is toxic. Use only under a fume hood. Avoid contact with skin. Wash thoroughly with soap and water. Use caution when venting the separatory funnel. Pressure may build up in the funnel. Point the end of the funnel toward the back of the fume hood.

Experimental - week one

Add 50.00 to 100.00 mL of the beverage (or the amount you determined above) and 25 mL of dichloromethane to a 250 mL separatory funnel. Stopper the separatory funnel, and gently invert and vent it by opening the stopcock. Swirl the contents in this inverted position for 5 to 10 minutes venting regularly. Do not shake. Be sure to use gentle motions to avoid formation of an emulsion. Return the separatory funnel to the upright position and allow the phases to separate.

When a distinct boundary forms, drain the dichloromethane layer and the emulsion into an Erlenmeyer flask until the boundary just enters the stopcock. Extract the aqueous layer in the separatory funnel with a second 25 mL portion of dichloromethane. Swirl and drain the dichloromethane and emulsion layer into the same flask as the first portion. Discard the aqueous layer and rinse the separatory funnel with water.

Filter the liquid through phase paper into a 50 mL preweighed beaker. Allow the solvent to evaporate to dryness under a fume hood. Do not heat the sample; caffeine can be lost due to sublimation. Determine the grams of caffeine in the sample during week 2.

Experimental - week two

The product collected is a crude sample. Weigh the sample before proceeding. Note its color and odor. To verify that the sample is caffeine, a melting point[8] of the sample should be compared to that of a known sample of pure caffeine. Before beginning, look up the melting point of caffeine in the CRC Handbook of Chemistry and Physics. A small amount of pure caffeine and each sample are placed in separate capillary tubes and packed in the end. Place the capillary tubes in the melting point apparatus, which the instructor will show you how to operate. When the samples begin to melt, note the temperature and any changes in color or texture of the sample. Once the samples have completely melted or sublimed, note the temperature again. The temperature and the range of temperatures at which melting occur are used to identify a substance and its purity. Comment on the identity of the substance extracted as well as its purity.

The following operations should be carried out UNDER THE FUME HOOD. Dissolve a small amount of the caffeine extracted in about 1 mL of dichloromethane. At the same time, dissolve an approximately equal amount of pure caffeine in a second 1 mL of dichloromethane. Obtain a thin layer chromatographic plate. About 1.5 cm from the bottom, mark the spotting line in pencil. DO NOT USE PEN. Next, spot each of the caffeine solutions by dipping the end of a clean capillary tube into the solution then gently touching the tube to the plate. Prepare the developing chamber which may be a jar with lid or a 1 L beaker with watch glass. The developing solvent is a 1:1 mixture of ethyl acetate:methanol. Be sure that the level of the solvent will not go over the spotting line on the plate. Allow the solvent to move up the plate until it is about 1.5 cm from the top of the plate. Remove the plate and allow it to air dry. Once it is dry, illuminate it with a UV light source. Mark the solvent front and any spots visible by circling with a pencil. Determine R_f values for all spots. Identify any caffeine in the beverages.

Disposal Methods

Flush excess beverage down the drain with water unless otherwise directed by the instructor. Ethyl acetate/methanol is placed in the non-halogenated organic waste.

Questions

1. What is the caffeine content of your beverage(s)?
2. What are the possible health effects of caffeine?
3. How does your caffeine content compare with literature values?
4. What is the melting point of caffeine? Of your product?
5. What is the R_f value for caffeine? For the components of your product?

Determination of Sugar as Glucose[9]†

This determination of sugar or carbohydrates in the beverage is based on the color which results when the sugars present reduce 3,5-dinitrosalicylic acid, DNSA, to 3-amino-5-nitrosalicylic acid according to the following reaction.

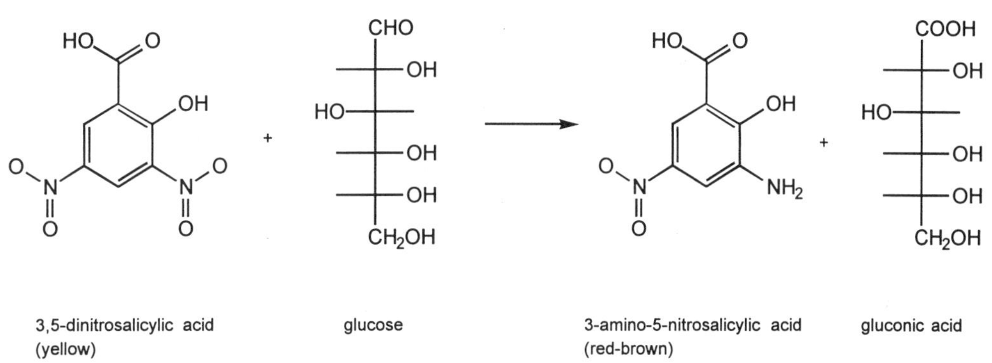

Complex sugars and carbohydrates are broken down into simple sugars in the first step when the sample is boiled in hydrochloric acid. The solution is then made basic and simple sugars are good reducing agents in basic solutions. The amount of DNSA reduced is directly proportional to the amount of sugar present.

Most beverages will have to be diluted by at least a factor of 10 (this is called a 10:100 dilution) before carrying out the analysis. However, you will probably want to make more than one dilution to be sure that you obtain more than one acceptable reading. The dilutions must be carried out quantitatively. Consider whether you expect your beverage to have a high or low concentration of sugar or carbohydrates. Some suggested dilutions include: 1:100, 2:100, 3:100, 5:100, 10:100. Most likely, you will not need to do all of these dilutions.

---

†Reprinted with permission from Laboratory Manual to Accompany Chemistry by A.D. Baker, L.F. Gries, M.H. Navidi, copyright West Publishing Company, 1990, pages 379-381.

This analysis is a colorimetric method and the color analyzed is time dependent. Therefore, once the 3,5-dinitrosalicylic acid is added, begin timing and make the absorbance measurements at a fixed time after the addition. The time should be between 5 and 20 minutes after addition.

A colorimetric analysis is based on Beer's law:

$$A = \epsilon l C$$

where A is the amount of light absorbed by a solution, ϵ is the molar absorptivity of the absorbing species, l is the pathlength of the light through the solution, and C is the concentration of the absorbing species. The molar absorptivity is a characteristic of the absorbing molecule and is a function of the wavelength of light being absorbed. Thus, at a fixed wavelength and fixed path length, the absorbance is directly proportional to the concentration of the analyte.

To use Beer's law, a calibration curve is prepared. The absorbances of solutions with known concentrations are measured and a plot of absorbance vs. concentration is made. Then by knowing the absorbance of a solution, its concentration can be found by reading the value from the calibration curve.

From the equation above, if the concentration is zero, then the absorbance should also be zero. Thus you will need a zero or blank sample to set the spectrometer to zero. The blank is prepared in the same way as the other samples and standards except that water is added in place of the sample. Thus any absorbance that is due to reagents other than the one you want to measure will be subtracted out.

Safety

CHECK FOR FLAMMABLES BEFORE LIGHTING A BUNSEN BURNER. When boiling solutions, point test tubes away from your face. Use test tube holders to handle hot glass. Strong acid and base burn skin. Wash all spills immediately and inform the instructor. The DNSA solution is TOXIC; avoid contact with the skin. Wash any spill immediately with soap and water. Inform the instructor.

Experimental

Turn on the spectrometer and set the wavelength to 580 nm. Prepare four standards and a blank at the same time. Stock solutions of glucose in water will be provided. The samples can also be prepared at the same time or at a later time.

Transfer 2.000 mL of each sucrose standard to a large test tube. Use 2.000 mL of water for the blank. Add 2.000 mL 6 M HCl to each test tube and place in a boiling water bath for 10 minutes. Remove the test tubes and carefully add 8.00 mL of 2.5 M NaOH, then 2.000 mL of 0.05 M 3,5-dinitrosalicylic acid to each test tube. As soon as the DNSA is added, cover with Parafilm and shake to mix solutions thoroughly. Place in a boiling water bath for 5 minutes. Begin timing. Cool the solutions in an ice-water bath for 10 minutes. Transfer each solution to a cuvette. Use the blank to zero the spectrometer. Record the absorbance of each standard and plot the calibration curve. Be sure it is linear before determining the concentration of the samples. Note that a Spectronic 20 is most accurate when the absorbance falls between 0.1 and 0.8. Therefore, the most accurate estimates of the concentration of the samples will occur when the absorbance is within this range.

Prepare at least 2 dilutions of your beverage in the range of 1:100 to 10:100, depending on the amount of sugar or carbohydrate you expect. Prepare the samples as above. Transfer 2.000 mL of the diluted beverage to a large test tube. Follow the instructions given above. Be sure to measure the absorbance at the same time interval from the addition of DNSA as used for the standards.

Disposal Methods

Place all solutions in properly labelled waste beakers or bottles. Do not flush DNSA down the drain.

Questions

1. What level of sugar or carbohydrate was found in the beverage(s)?
2. How do your values compare with the literature values?

Alcohol Content by Distillation[2-4]

Distillation is a method of separating liquids based on a difference in their boiling points. Ethanol, the alcohol in beverages, boils at 78.5 °C while water boils at 100 °C. Therefore, in theory, by carefully heating a mixture of ethanol and water, one can boil off the ethanol leaving the water behind. However, in the case of an ethanol-water mixture, complete separation by distillation is not possible because the liquids form an azeotrope. Nevertheless, the amount of alcohol in a sample can be estimated by distillation. The sample to be analyzed is heated slowly to about 98 to 100 °C. From the total volume of distillate collected and the density of the distillate, the amount of ethanol can be calculated. The D pages of the CRC Handbook has tables of the specific gravity of mixtures of ethyl alcohol and water by volume and by weight. Once the percent of ethanol in the distillate is known, the total volume of distillate is used to calculate how much ethanol was in the sample. Assume the distillation completely removes the ethanol from the sample; i.e., the amount of ethanol collected is the same as the amount in the beverage sample. Therefore, the percent alcohol by volume in the sample can be calculated.

The presence of alcohol in the distillate must be confirmed. Ethanol reacts with dichromate in sulfuric acid solution to yield Cr^{3+} and acetic acid. This reaction is used in breathalyzer tests of sobriety. Balance the following reaction:

$$\underset{\text{orange}}{Cr_2O_7^{2-}} + CH_3CH_2OH \rightarrow \underset{\text{green}}{Cr^{3+}} + CH_3COOH$$

Safety

Safety glasses must always be worn in the lab. Use caution when handling glassware which may be hot. Chromium is toxic so avoid contact with skin. The dichromate solution is strongly acidic. Use caution when adding it to the acetone and when adding the distillate to the solution, because splattering may occur. Place chromium waste in labelled bottles or beakers.

Experimental

Set up the distillation apparatus as in Figure 1. To a 125 mL round bottom flask, add 50.00 mL of the beverage, 25.00 mL of water and 1-2 boiling stones. Slowly heat the sample until the temperature of the vapor reaches 98 to 100 °C. Determine the total volume of distillate and its density. Since water and alcohol form an azeotrope, a complete separation of the alcohol is not possible. Density tables in the CRC Handbook of Chemistry and Physics can be used to determine the volume of alcohol in the distillate based on the density.

The presence of ethanol in the distillate should be confirmed by the chromic acid test. Add a drop of chromic acid reagent to 1 mL of acetone in a small test tube and observe the color. If the solution turns green, do not use the acetone. Instead, get a new sample. Add 1-3 drops of the distillate to 1 mL of acetone then add 1 drop of the chromic acid reagent. Note the color. A green solution is a positive test for the presence of alcohol.

Disposal Methods

Flush distillate down the drain with water. Place solutions used for the chromic acid test in properly labeled bottles or beakers.

Questions

1. What is the alcohol content of the beverage?
2. How does your value compare with the manufacturer's value?

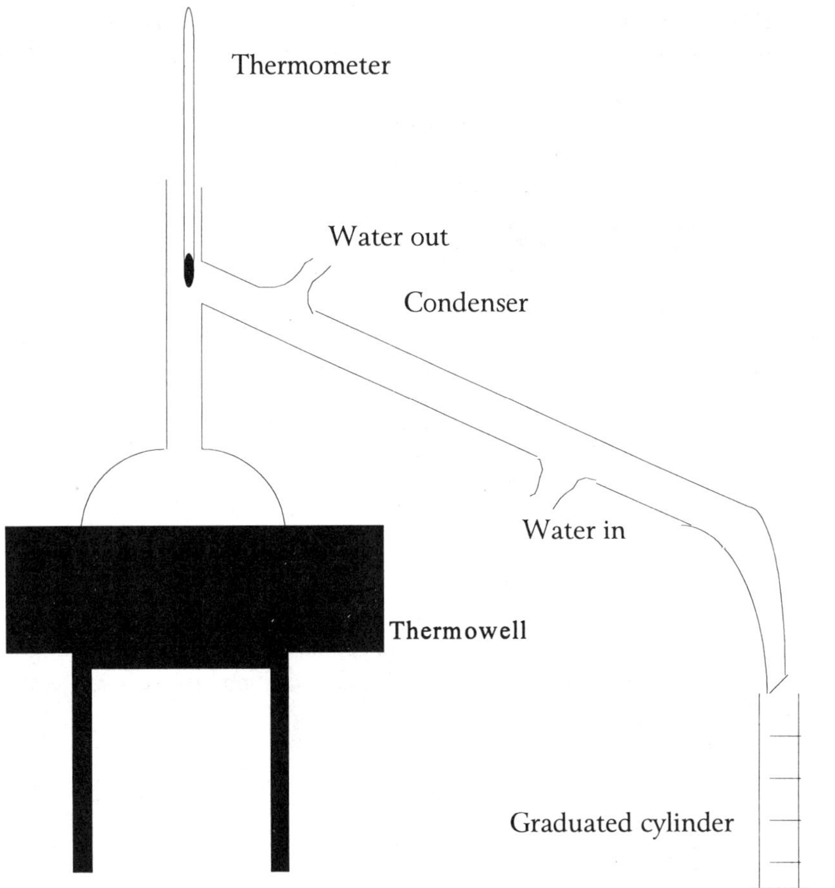

Figure 1. Distillation set-up for the alcohol content of a beverage.

Determination of the Percent Solids in Beverages

The percent solids present in beverages such as orange juice is regulated by law. The test for percent solids can be run simultaneously with other experiments. Be sure a literature value exists for your beverage or the manufacturer can provide data for comparison.

Safety

Eye protection must always be worn in the lab. Use caution when handling hot equipment.

Experimental

Label and dry aluminum pans at 110 °C for 1 hour. Cool for 15 minutes in a desiccator and determine the mass of each. Dry, cool and determine the mass of the pans until the mass changes by no more than 0.5 mg between cycles. Note that subsequent heating cycles may be reduced to 30 minutes.

Measure out an accurately known amount of the beverage and transfer it to an aluminum weighing pan. Dry to constant mass.

Questions

1. Consider whether you expect your beverage to have a significant amount of solids. What solids may be present?

Disposal Methods

Place used pans in the trash unless otherwise directed by the instructor.

References

1. Experience the Extraordinary Chemistry of Ordinary Things, T. Chasteen and B. Richardson, John Wiley and Sons, Inc. 1992, p.139.
2. Experience the Extraordinary Chemistry of Ordinary Things, T. Chasteen and B. Richardson, John Wiley and Sons, Inc. 1992, p.95.
3. The Organic Chem Lab Survival Manual, 3rd Ed., J.W. Zubrick, John Wiley And Sons, Inc., 1992, p.187.
4. The Systematic Identification of Organic Compounds, 6th Ed., R.L. Shriner, R. C. Fuson, D.Y. Curtin, T. C. Morrill, John Wiley and Sons, Inc., 1980, p. 149
5. Organic Chemistry Experiments, B.N. Campbell, Jr., M. McCarthy-Ali, Brooks/Cole Publishing, 1994, p. 147.
6. Experience the Extraordinary Chemistry of Ordinary Things, T. Chasteen and B. Richardson, John Wiley and Sons, Inc. 1992, p.183.
7. The Organic Chem Lab Survival Manual, 3rd Ed., J.W. Zubrick, John Wiley And Sons, Inc., 1992, p.143.
8. The Organic Chem Lab Survival Manual, 3rd Ed., J.W. Zubrick, John Wiley And Sons, Inc., 1992, p.243.
9. Laboratory Manual to Accompany Chemistry, A.D. Baker, L.F. Gries, M.H. Navidi, West Publishing Company, 1990, 379.
10. The Organic Chem Lab Survival Manual, 3rd Ed., J.W. Zubrick, John Wiley And Sons, Inc., 1992, p.127.

Preparation
Beverages

Name ____________
Section____________
Date ____________

Possible beverages and experiments are listed in the **Experimental** section for this experiment. You may study other beverages or carry out other experiments with the instructor's approval. Write a proposal of the experiments you plan to carry out and how they will answer the questions in the proposal. Specify what each group member will do.

Practice problems

1. A student followed the procedure for titrating vitamin C in a juice. She recorded the following data in her notebook. Answer the questions which follow.
Standardized vitamin C was titrated with 2,6-dichloroindophenol to standardize the indophenol. Aliquots of 15.00 mL were titrated to a rose pink endpoint. The standardized vitamin C was prepared by diluting 0.2801 g of ascorbic acid in a total volume of 1.000 L.

Titration Number	Final volume Indophenol	Initial volume Indophenol	Total volume Indophenol
1	21.22	10.53	
2	31.93	21.22	
3	42.61	31.93	

Titration of juice (50.00 mL of juice diluted to 100.00 mL with metaphosphoric acid/acetic acid reagent). Aliquots of 15.00 mL of juice solution were titrated with indophenol solution to a rose pink endpoint.

Titration Number	Final volume Indophenol	Initial volume Indophenol	Total volume Indophenol
1	21.01	5.60	
2	36.46	21.01	
3	16.94	1.56	

A 1:1 dilution of the metaphosphoric acid/acetic acid solution was titrated for the blank. Aliquots of 15.00 mL of blank were titrated to the rose pink endpoint.

Titration Number	Final volume Indophenol	Initial volume Indophenol	Total volume Indophenol
1	10.79	10.58	
2	11.02	10.79	
3	11.22	11.02	

a.) Calculate the concentration of the standard vitamin C in mg/mL.
b.) What is the concentration of the indophenol in mg vitamin C/mL of indophenol?

c.) What is the concentration of the dilute juice sample in mg vitamin C/mL?
d.) What is the concentration of vitamin C in the original juice?
e.) How many ounces of juice provide the RDA of vitamin C?

2. A 50.00 mL sample of liquor was distilled. A total of 28.04 mL of distillate was collected. A 10.00 mL aliquot of distillate had a mass of 9.0631 g. What is the percent of alcohol in the distillate? What is the percent of alcohol in the original sample of liquor?

3. A sample of espresso was analyzed for caffeine. A 20.00 mL aliquot was extracted with dichloromethane as outlined in the procedure. When the solvent evaporated, 82.1 mg of white product remained. Further analysis of the product included a melting point (melting point range 216 to 225 °C) and a thin layer chromatogram. The TLC plates are shown at right. Calculate R_f values for caffeine and the components in the sample. Based on this data, is caffeine present in the espresso? Is the caffeine extracted pure? Justify your answers.

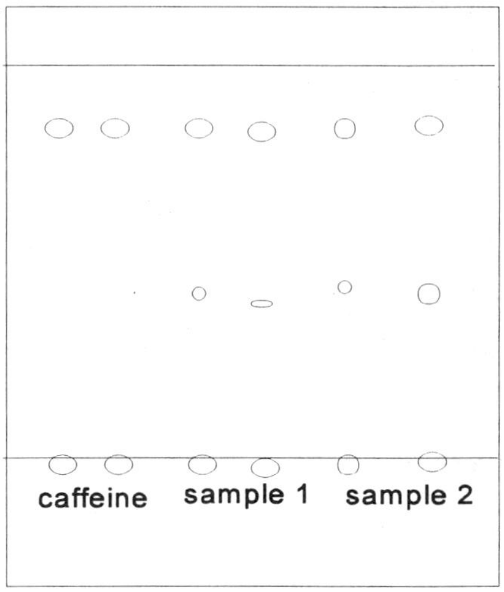

Assuming that the 82.1 mg extracted is pure caffeine, how much caffeine is present in a serving of espresso? (A serving is 1/4 cup).

4. A sample of beer was to be analyzed for total carbohydrates. A 10.00 mL sample was diluted to 100.0 mL in a volumetric flask. Next, samples of standard glucose were prepared as described in the **Experimental** section and the following calibration data obtained.

Concentration of glucose g/L	Absorbance at 580 nm
0	0
0.5004	0.10
1.0021	0.28
1.5002	0.40
1.9987	0.54

Construct a calibration curve based on this data.

The beer sample was prepared next and the absorbance was found to be 0.22. Based on this absorbance, what is the concentration in g/L of carbohydrates in the diluted beverage? What is the concentration of carbohydrates in the undiluted beer in g/L? How many grams of carbohydrates are in a serving of this beer?

5. The juice analyzed in question 1 was titrated to a phenolphthalein endpoint to determine the total acidity. A 50.00 mL aliquot of undiluted juice was titrated with 39.88 mL of 0.008512 M NaOH. What is the total acidity?

The only two acids present are ascorbic and citric. At the endpoint, the mole ratio of ascorbic acid to NaOH is 1 to 1 while the mole ratio of citric acid to NaOH is 1 to 3. What is the concentration of ascorbic acid in the juice? What is the concentration of citric acid in the juice?

Results
Beverages

Name ____________________
Section ____________________
Coworkers ____________________
____
____
____
Date ____________________

Experiment: Determination of ______________ in ______________.

Record data and observations in an appropriate format.

Beverages -- Oral Report Guidelines

Each member must present part of the report and the group will submit a brief (1-2 page) summary of the project at the oral presentation.
Limit the presentation to approximately 5 minutes per person.
Practice the presentation.
Overhead transparencies are available from the instructor with prior arrangements.

Include the basic parts of a report:

Introduction - what beverages you are studying and why. Also include some background on the experimental theory.

Experimental - a very brief description of the procedure.

Results - prepare clear charts or graphs to show results of your experiments. Include literature or manufacturer values for comparison to experimental results. Be prepared to discuss accuracy and precision for all of your results.

Conclusions - summarize what you learned. Propose further experiments or studies.

EVALUATION FORM
BEVERAGES

NAME:_________________ DATE:_____

Were you project director? ____ YES ____ NO

1. List the experiments that you did:

2. Using a number from 1 to 10, what grade would you give yourself? Justify your answer.

3. Using a number from 1 to 10, what grade would you give each of your partners. Give examples from lab, outside meetings, and contributions to the report to justify your evaluation.

4. What did the lab instructor do that was helpful/unhelpful in completing your project?

Buffers

4 lab periods; work in groups. Complete the Preparation pages for each week before laboratory.

Goals

- To determine the pK_a of acetic acid
- To determine the concentration of acetic acid in vinegar
- To prepare buffers of assigned pH and study their properties
- To use a qualitative analysis scheme to draw conclusions about buffers
- To quantitatively test conclusions about buffers
- To determine the buffer capacity of assigned buffers

Background

Buffer solutions are important to a wide range of chemical and biological systems. For example, buffers are responsible for maintaining physiological pH and for maintaining reaction conditions during industrial processes. A buffer is a solution that resists changes in pH when a strong acid or a strong base is added to the solution. Many conditions determine how effective a buffer solution is under given conditions. For example, the total concentration of the buffer, the moles of strong acid or strong base added to the buffer, the conjugate acid/base pair used to prepare the buffer and the pH of the solution all contribute to the effectiveness of a buffer solution.

Acetic acid is an organic acid found in ordinary vinegar. Acetic acid and its conjugate base, acetate ion, are used to prepare buffers. Baking soda, or sodium bicarbonate, is a leavening agent used in foods. When baking soda is mixed with an acidic ingredient such as milk or is heated, carbon dioxide is released causing cakes and other baked goods to rise. The bicarbonate ion acts as a weak acid and with its conjugate base, carbonate ion, is also used as a buffer.

In Week One, a stock solution of 0.1 M NaOH is prepared and standardized against potassium hydrogen phthalate, KHP. Sodium hydroxide is hygroscopic and absorbs water from the air. Thus, weighing an exact amount of solid to prepare a solution of known molarity is not possible. Therefore, we make a solution with approximately the molarity we want and then titrate the prepared solution against a primary standard. The primary standard in this experiment is potassium hydrogen phthalate, KHP. The primary standard is a material that is available in a very pure, dry state whose molar mass is known with precision. Also, its reaction with the solution being standardized must be quantitative with known stoichiometry as shown below:

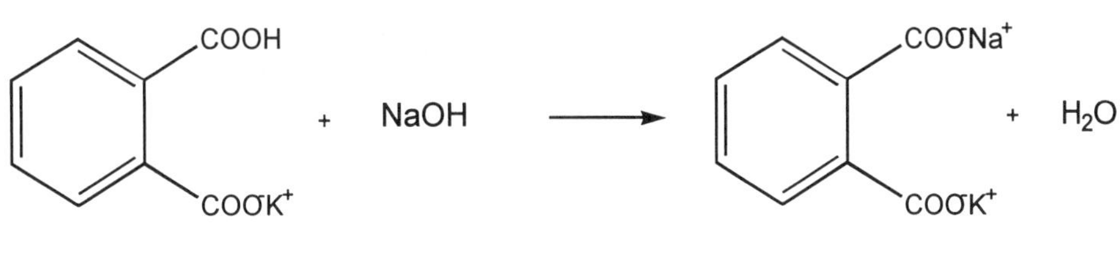

KHP

The reaction of KHP with sodium hydroxide

While the design of the Erlenmeyer flask is ideal for indicator titrations, pH titrations are carried out in a beaker. With an indicator, only the tip of the buret must fit in the mouth of the Erlenmeyer flask. The narrow opening permits swirling the contents without spilling the solution. Also, the sloped sides allow stray solution to be easily washed down with a wash bottle. However, when we follow a titration with a pH electrode and meter, we use a 150 mL beaker. The opening of the beaker is large enough to allow the pH electrode to remain in solution while the buret is also arranged over the opening. The walls are tall enough so that solution is not splashed out by the magnetic stirrer. A typical arrangement is shown in Figure 1.

To standardize the NaOH, you must know the exact number of moles of KHP being titrated. Two methods are available to transfer an exact number of moles of KHP to be titrated. The first method is to weigh out a known amount of KHP by difference into the beaker or Erlenmeyer flask. The solid is then dissolved in distilled water and the titration completed. A second method is to prepare a solution of KHP with a known molarity and then to quantitatively transfer the solution to be titrated. While either method works well, the former generally involves less error and is the method of choice for analytical chemists. However, the second method is suitable for work of this nature.

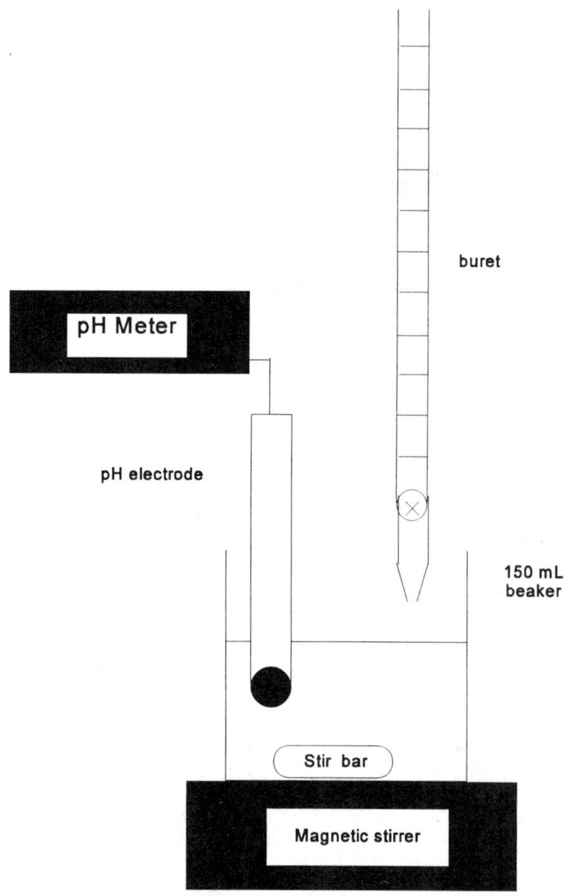

Figure 1. Experimental arrangement of a pH titration.

In Week Two, the purity of a commercial product, vinegar, is determined. The pK_a of the weak acid is also found. The standardized NaOH previously prepared is used to titrate vinegar using a pH meter to follow the titrations. A pH titration curve is used to find both the purity and pK_a. From the pH titration curve, appropriate indicators are selected for indicator titrations. A comparison is made between indicator and pH titrations. Vinegar is approximately 5% (wt/vol) of acetic acid in water. Before titrating, dilute the vinegar to an appropriate concentration as discussed in the preparation.

The class will discuss pH titration curves. Differences between titrating strong and weak acids will be noted. The class will find the average pK_a of acetic acid. Discussion will include the accuracy and precision of the data.

For Week Three, you will be assigned buffers to prepare and test using a qualitative analysis scheme. The buffers will be prepared from commercial vinegar. Consider how you will prepare the buffer at the assigned pH recalling that the total concentration of a buffer is the sum of the concentrations of the weak acid and its conjugate base. Therefore, in a 0.1 M buffer, the concentration of acetic acid plus the concentration of acetate ion is 0.1 M.

From the results of Week Three, you will make predictions about buffers. The class will compile results and use all of the results in making predictions. Then in Week Four, you will test the predictions by making quantitative measurements of the buffer capacity of the buffers you prepared.

What conditions are necessary to prepare a good buffer?
What factors determine buffer capacity?

Equipment and Materials

pH meter with electrode
magnetic stirrer with stir bar
plastic storage bottles 1 L and 250 mL
25, 50 mL volumetric pipets
250, 500 mL volumetric flasks
Erlenmeyer flasks 125 or 250 mL
50 mL burets
16 X 125 mm test tubes
10 mL graduated cylinders
pH indicator strips
standard buffers, pH 4, 7,and 10
indicators: phenolphthalein, bromothymol blue, thymolphthalein, bromocresol green
potassium hydrogen phthalate (KHP) dried
vinegar
14 M NaOH

Week One

Safety

Eye protection must always be worn in the lab. Sodium hydroxide poses a special threat to the eyes! Use caution. The sodium hydroxide stock solution is very corrosive. Wear gloves and safety glasses when working with this material. Rinse all glassware with water after use to prevent accidents and to avoid etching the glassware. When adding base to a buret, lower the buret and use a funnel. POURING SOLUTIONS ABOVE EYE LEVEL IS VERY DANGEROUS. Rinse the buret, including the tip, with distilled water when finished with the titrations. Fill the buret with distilled water and cap to store.

Experimental

Record all data in your notebook or on the **Results** page, as directed. Prepare 1 L of approximately 0.1 M NaOH from the stock solution of 14 M NaOH. Store this solution in a 1 L plastic bottle labeled with your name, date and lab section. Prepare your KHP for titration either by weighing by difference the amount to be used in the titration or by preparing a solution of known molarity.

Select an appropriate indicator from the list above for this titration of a weak acid with a strong base. Find the average molarity of the NaOH solution you prepared. Estimate the precision of your titrations by calculating the standard deviation. The relative error in the data is found by dividing the standard deviation by the average. The relative error, SD/average, should be less than 0.004. If your error is larger than 0.004, see your instructor.

Clean the buret and fill it with distilled water. Save the NaOH for Week Two.

Disposal Methods

All solutions, except the standardized NaOH, can be washed down the drain with plenty of water, unless otherwise directed by the instructor.

Questions

1. Report the average molarity of your NaOH.
2. Report the precision of your data with a standard deviation.
3. Could you use HCl to standardize NaOH? Why or why not?

Week Two

Safety

Eye protection must always be worn in the lab. Sodium hydroxide poses a special threat to the eyes! Use caution when handling NaOH. Clean up any spills immediately. Lower the buret to fill with NaOH. Rinse all glassware with water when done. Rinse the buret, including the tip, with distilled water. Fill the buret with distilled water when done.

Experimental - Part 1 To be done with a partner

Record all data in your notebook or on the **Results** page, as directed. Prepare a 500.00 mL dilution of commercial vinegar according to your calculations on the **Preparation** page. Prepare the pH meter according to your instructor's directions. Carry out the pH titration. Titrate 25.00 mL of the vinegar solution you prepared with your standardized 0.1 M NaOH using a pH meter to record pH after every 0.5 to 1 mL of titrant added. When the pH begins to change rapidly, reduce the volume of titrant added. Titrate to a constant pH approximately equal to the pH of the titrant. Plot pH vs. the volume of titrant added. Find the equivalence point as described in **Finding the Equivalence Point of a pH Titration Curve** in this manual. Find the pK_a for acetic acid using your knowledge of the Henderson-Hasselbach equation. Repeat the titration at least once more. Plot a rough sketch of the titration curve. From this sketch, determine an appropriate indicator to use to carry out this titration.

Experimental - Part 2 To be done on your own

Use an indicator to titrate the vinegar and find the molarity of the diluted vinegar solution. See Week One for guidelines on calculating the precision of your titrations.

You will be assigned buffers to prepare for next week. Save the standardized NaOH.

Disposal Methods

All solutions, except the standardized NaOH, are flushed down the drain with water, unless otherwise directed by your instructor.

Questions

1. Prepare a titration curve from the data. Report the pH at the equivalence point and justify your choice of indicator.
2. Report the pK_a for acetic acid from your data. Compare your value to the class average.
3. Report the molarity of acetic acid in the diluted sample from **Part 1** and **Part 2**. Include an estimate of precision for each method.
4. Report the molarity of acetic acid in the undiluted vinegar.
5. Report the weight/volume percent of acetic acid in vinegar.
6. How are pH and indicator titrations similar? How is each unique?

Week Three

Safety

Eye protection must always be worn in the laboratory. Use caution when handling the acid and base used in this part of the experiment. Clean up any spill immediately. Rinse all glassware with water when done.

Experimental

Record all data in your notebook or on the **Results** page, as directed. Prepare 500.00 mL of a 0.1 M buffer at your assigned pH. Start with commercial vinegar. Also, prepare 250.00 mL of a 1:2 dilution of the buffer. Since you are starting with only one half of the buffer, you must adjust the pH before diluting to volume.

Test each buffer against strong acid and strong base. Add 3 mL of the undiluted buffer to seven test tubes. The first test tube serves as a blank. Next, add 0.1 M NaOH to the remaining six test tubes such that a different amount of base is added to each: 0.5, 1, 1.5, 2, 2.5, or 3 mL. Mix the solutions. Test each solution with pH paper. Record the volume of base added and the final pH of the solution. Repeat the experiment with the diluted buffer.

Next, test the buffers against 0.1 M HCl. Add 3 mL of the undiluted buffer to six test tubes. The blanks from the first part of this experiment can be reused for this part. Next, add 0.1 M HCl to the remaining test tubes such that a different amount of acid is added to each: 0.5, 1, 1.5, 2, 2.5, or 3 mL. Mix the solutions. Test each solution with pH paper. Record the volume of acid added and the pH of the solution. Repeat the experiment with the diluted buffer.

Record the class results and wait for the discussion. Save the buffer solutions for the next lab.

Disposal Methods

Wash all solutions, except the buffers, down the drain with water unless otherwise directed by the instructor. The pH paper may be thrown in the trash.

Questions

1. Compare your data with the class data. Which buffer most strongly resists the addition of base? Rank the buffers according to their ability to resist addition of base. How do the pH of the original buffer and the pH after addition of base relate to the pK_a for each of these solutions?
2. Which buffer most strongly resists the addition of acid? Rank the buffers according to the ability to

resist addition of acid. How do the pH of the original buffer and the pH after addition of acid relate to the pK_a for each of these solutions?
3. Based on these results, which buffer do you expect to have the greatest buffer capacity toward NaOH? toward HCl? Which buffer do you expect to be the best buffer? Why?
4. Explain the effect of dilution on the buffer's ability to resist a strong acid or base.

Week Four

Safety

Eye protection must always be worn in the laboratory. Use caution when handling acids and bases. Lower the buret to fill it with acid or base. Wash any spills immediately with plenty of water. Rinse the buret, including the tip, with distilled water and fill to store.

Experimental

Record all data in your notebook or on the **Results** page, as directed. Based on the results of the previous week, you made predictions about the buffer capacity of the various buffers you and your classmates prepared. Recall that the minimum volume for a good titration is 10.00 mL. Test your buffers, diluted and undiluted, toward the strong acid HCl and the strong base NaOH. Use your standardized NaOH. If you are out of standardized solution, NaOH will be provided. Calculate a buffer capacity for each buffer. Compile the class data and wait for the discussion.

Disposal Methods

Flush all solutions down the drain with water unless otherwise directed by the instructor.

Questions

1. What is the buffer capacity of your buffers to strong acid and strong base?
2. Do the buffer capacities agree with the predictions made in Week Three? Why or why not?
3. What is the effect of dilution on the buffer capacity? Quantify your answer.
4. What is the relationship between pH, pK_a, and buffer capacity for a strong acid? For a strong base?
5. Can the acetic acid/acetate system be used to prepare a buffer of pH 9.2? Why or why not?
6. Can the acetic acid/acetate system be used to prepare a buffer of pH 5.0? Why or why not? Use the titration curve to justify your answer.

Preparation
Buffers - Week One

Name _______________
Section _______________
Date _______________

Read

1. The **Background** and **Week One** of this experiment.
2. **Use of Volumetric Glassware** in this manual.
3. About indicators and titrations in your text.

Practice Problems

1. Calculate the volume of 14 M NaOH needed to prepare 1 L of 0.1 M NaOH. Describe how you would prepare the solution, including any glassware necessary.

2. Calculate the amount of KHP needed for each titration assuming the NaOH is 0.1 M and a titration should use 20 to 30 mL of titrant.

3. Assume that you want to prepare a solution of KHP to titrate as your standard. Describe in detail how you would prepare the solution.

Preparation
Buffers - Week Two

Name ______________
Section ______________
Date ______________

Read

1. The **Background** and **Week Two** of this experiment.
2. **Finding the Equivalence Point on a pH Titration Curve** in this manual.
3. About indicators and titrations in your text.

Questions

Define the following terms:
Equivalence point
Half-neutralization point

Write the Henderson-Hasselbalch equation.

Practice Problems

1. Commercial vinegar is 5 % acetic acid (wt/vol). Calculate the molarity of acetic acid in vinegar.

2. Based on the fact that a titration should use 20 to 30 mL of titrant, calculate approximately how many moles of your standardized NaOH will be used per titration.

3. What volume of commercial vinegar has an equivalent number of moles to the volume of NaOH used in 2 above. Is this a reasonable volume to titrate? Why or why not?

Preparation
Buffers - Week Three

Name ______________
Section ______________
Date ______________

Read

1. The **Background** and **Week Three** of this experiment.
2. About buffers in your text.

Questions

1. Describe how you will prepare your buffer from commercial vinegar and the supplies listed in this experiment. Include calculations to show the amounts of reagents needed. Recall that the concentration of a buffer is the total concentration of the conjugate acid and base.

2. Describe how you will prepare 250.00 mL of a 1:2 dilution of the buffers. Include all glassware needed. What is the new molarity of the buffer?

Preparation
Buffers - Week Four

Name ____________
Section ____________
Date ____________

Read

1. The **Background** and **Week Four** of this experiment.
2. Review assignment and data from Week Three.

Questions

1. Describe how to measure the buffer capacity to a strong acid. Be sure to include the glassware and solutions needed to make the measurement.

2. For one of the buffers prepared in **Week Three**, estimate the buffer capacity. Show all calculations.

Results
Buffers - Week One

Name _______________
Coworker(s)_______________
____
Section _______________
Date _______________

Data

Preparation of NaOH

Volume of 14 M NaOH used ______
Total volume of solution ______
New molarity of NaOH ______

Table 1. Standardizing NaOH with KHP

Trial	KHP used	final vol (mL)	initial vol (mL)	total vol (mL)	comment	molarity of NaOH

Average molarity of NaOH __________
Standard deviation __________
Relative error __________

Answers to Questions

Results
Buffers - Week Two

Name ________________
Coworker(s) ________________
____
Section ________________
Date ________________

Data

Preparation of Diluted Vinegar

Volume of vinegar used ______
Total volume ______

pH Titration of Diluted Vinegar

Table 1. Volume of NaOH used - pH data for pH titrations

Trial 1			Trial 2		
Volume NaOH (mL)	pH	Total volume NaOH (mL)	Volume NaOH (mL)	pH	Total volume NaOH (mL)

Indicator titrations of diluted vinegar

Table 2. Titration of vinegar with NaOH

Trial	vol of vinegar (mL)	final vol (mL)	initial vol (mL)	total vol (mL)	comment	molarity of vinegar

Average molarity of vinegar ________
Standard deviation ________
Indicator ________

Answers to Questions

Results
Buffers - Week Three

Name ____________
Coworker(s) ____________
____
Section ____________
Date ____________

Data

Preparation of Buffers

Buffer pH = ______ Volume vinegar used ______

Dilution Volume of buffer ______ Total volume of solution ______

Testing Buffers with NaOH

Table 1. pH of buffer solutions after mixing with 0.1 M NaOH

Solution	Volume of 0.1 M NaOH added						
	0	0.5	1.0	1.5	2.0	2.5	3.0
Buffer							
Dilution							

Testing Buffers with HCl

Table 2. pH of buffer solutions after mixing with 0.1 M HCl

Solution	Volume of 0.1 M HCl added						
	0	0.5	1.0	1.5	2.0	2.5	3.0
Buffer							
Dilution							

Ranking of buffer resistance to addition of 0.1 M NaOH: ____________________

Ranking of buffer resistance to addition of 0.1 M Hcl: ____________________

Answers to Questions

Results
Buffers - Week Four

Name ______________
Coworker(s) ______________
____
Section ______________
Date ______________

Data

Determination of Buffer Capacity

Table 1. Buffer capacity against standardized NaOH

Buffer	Final vol (mL)	Initial vol (mL)	Total vol (mL)	Buffer capacity (mol/L)
Buffer				
Dilution				

Concentration of NaOH used ______________

Table 2. Buffer capacity against standardized HCl

Buffer	Final vol (mL)	Initial vol (mL)	Total vol (mL)	Buffer capacity (mol/L)
Buffer				
Dilution				

Concentration of HCl used___________

Answers to Questions

Carbonates and Hydrogen Carbonates

2 lab periods; work in groups. Complete the Preparation page before laboratory.

Goals

- To identify an unknown carbonate or hydrogen carbonate of a Group 1 or 2 cation
- To prepare a written report identifying the unknown

Background

Seek help before your lab period if you have questions about this assignment. The procedures given in the **Experimental** section are guidelines for you to follow but you should expect that during your experimentation the amounts of reagents or procedures will have to be modified. You should have ample time to repeat experiments to get the best accuracy and precision possible. All group members should be able to carry out the calculations for each experiment.

When trying to identify an unknown compound, frequently the unknown material is dried first. Your instructor may issue your unknown a week before this assignment so you can dry your sample. Place the unknown in an open weighing bottle in a drying oven labeled with your name, section, and unknown code. Alternatively, the unknown may be issued to you dried. Either way, keep the dried sample in a desiccator to prevent it from absorbing water from the atmosphere.

Each group will be given a single unknown, which is a Group 1 or 2 carbonate or hydrogen carbonate. From the experiments listed below, devise a plan to identify the unknown. Once it is identified, if time permits select experiments to confirm the identity of your unknown. Note that the goal of most of the experiments is to determine the molar mass for your unknown. The various experiments have different sources of error and thus different accuracy and precision. You need not do every experiment to confirm the identity of the unknown, but you are responsible for understanding the chemistry of each procedure. Note that the experiments are not listed in any particular order. Be sure to write the necessary chemical equations for each procedure in your notebook or on the **Results** page before carrying out the experiment.

Remember that this is an experiment. While approximate amounts of reagents can be suggested for each of the following procedures, you must determine the appropriate amounts for your unknown. For example, 0.5 g of $LiHCO_3$ will yield more moles of reactant than will 0.5 g of $BaCO_3$. In addition to the unknown, you will have available a known carbonate and hydrogen carbonate for comparison. This will allow you to try out an experiment, see the results, and carry out the calculations. From this information, you can estimate the accuracy expected for the particular experiment.

The most familiar hydrogen carbonate is the sodium salt found in most kitchens. Baking soda is used instead of yeast in quick breads and other baking. Usually an acidic ingredient is added to help release CO_2. However, heating baking soda generates CO_2 also. Sodium hydrogen carbonate is a base also used as a cleaning powder.

Calcium carbonate is found in the shells of marine animals such as coral, clams, and oysters. It is also found in limestone formations. One common use of this particular carbonate is as the active ingredient in antacids, such as Tums® and Rolaids®, and in calcium supplements for women.

What types of chemical tests are used to distinguish between a carbonate and a hydrogen carbonate? How do you identify an unknown chemical?

Equipment and Materials

eudiometer
drying oven
crucibles
50 mL burets
pH meter
desiccator
balances
concentrated HCl
phenolphthalein
acetone
methyl red
ethanol
LiCl
$CaCl_2$
$MgCl_2$
KCl
6 M HCl
6 M NaOH
0.1 M NaOH (standardized)
0.1 M HCl (standardized)
$NaCO_3$
$NaHCO_3$
unknowns

Mass Loss[1]

This experiment is based on the mass lost by your compound when it is reacted with acid, HCl. The goal is to find a molar mass for the unknown. Before proceeding, write a net ionic equation for the reaction of a carbonate or a hydrogen carbonate with HCl. What is being lost? Consider how many experiments will be needed for good accuracy and precision. NOTE: OPEN CONTAINERS ARE NOT ALLOWED ON THE ANALYTICAL BALANCES!

Before beginning, consider the goal of this experiment which is to find the molar mass of the unknown from the mass lost when the reactants are mixed. What data do you need? Remember that matter cannot be created or destroyed so the mass of the reactants must equal the mass of the products. How will you measure the mass of the reactants? the products? Carry out a trial run. This trial can be run with a known. From the data, calculate the number of moles of reactants you started with by calculating the number of moles of gas given off. Be sure that you are recording all of the data you need before proceeding. If you used a known compound, estimate the accuracy and consider what changes you could make to improve the results.

Safety

Eye protection must always be worn in the laboratory. The acid and base used in this experiment may irritate the skin. Wash any spills immediately with water. Reactions can be quite vigorous, so use caution when mixing reagents.

Experimental

Weigh approximately 2 g of unknown by difference to the nearest 0.1 mg. Note that you need to know exactly how much unknown is actually in the reaction vessel. Add approximately 20 mL of 6 M HCl to the unknown. You will need to weigh by difference the actual amount of HCl added. (See NOTE above) Agitate the solution and allow five minutes for the reaction to go to completion. Weigh the final solution. From the data you recorded, calculate the total mass of the reactants and the total mass of the solution after the reaction is complete. From the mass of the reactants and the remaining products, calculate the mass of gas lost and the moles of gas produced. Calculate the molar mass of the unknown. From the data estimate the accuracy and precision of the results. Do you need to modify the procedure or the amounts of reactants? Did you record all of the data you need to calculate a molar mass? If not, what data are missing?

Disposal Methods

Solutions may be flushed down the drain unless otherwise directed by your instructor. Keep unused unknown in the desiccator.

Questions

1. What is the unknown based on this data?
2. How precise are the data? If you used a known, how accurate are the data?

Gravimetric Analysis[2]

Gravimetric analysis is an analytical technique that uses the mass of a product formed to determine the number of moles of an ion in a compound. Here it is used to determine the mass of metal in a weighed sample of the unknown carbonate or hydrogen carbonate. Before beginning, write a balanced equation for the reaction of the unknown with HCl. What is the product containing the metal cation that is being isolated here? What products of this reaction are lost when the crucible is heated? Consider how many trials will be needed for accuracy and precision.

When carrying out gravimetric analyses the products isolated must be pure and dry. Therefore, the product is dried to constant mass. The product is dried, the mass found and the product dried and weighed again. This procedure is repeated until the mass changes by less than 0.5 mg between dryings. Other considerations include using the same balance to weigh the samples, cooling the samples in a desiccator so the sample does not absorb moisture on cooling, and avoiding adding finger prints to the crucibles that can add mass.

Safety

Eye protection must always be worn in the laboratory. Handle concentrated HCl with care. Wear gloves and use under a fume hood. Wash any spills immediately with water. Reactions can be quite vigorous, so use caution when mixing reagents.

Experimental[2]

Prepare crucibles and covers by washing and drying to constant mass over a Bunsen burner flame for 5 minutes or at 110° C in a drying oven for one hour. Cool in a desiccator for at least 15 minutes before weighing. Weigh each crucible and cover to the nearest 0.1 mg, taking care to note any distinguishing features on the crucibles. Heat again for 5 minutes over a Bunsen burner flame or 30 minutes in a drying oven, cool and reweigh to be sure the crucibles are at constant mass.

Add approximately 1 g of unknown to a preweighed crucible. Determine the mass of the unknown in the crucible to 0.1 mg. Under a hood, carefully add concentrated HCl dropwise until gas formation ceases. Then, add two more drops to ensure an excess of HCl. Partially cover the crucible to avoid splattering of the product. Slowly dry the product to constant mass over a Bunsen burner. Check that the proper data are being recorded by carrying out a sample calculation. If a known sample is used, find the accuracy of the method for determining the molar mass of the carbonate or hydrogen carbonate.

Disposal Methods

Flush all solutions down the drain with water unless otherwise directed by the instructor. The solid collected is flushed down the drain with water.

Questions

1. What is the unknown based on these data?
2. Estimate the precision of the method. If you used a known, estimate the accuracy.

Titrations[1]

The reaction of carbonates and hydrogen carbonates with standard HCl is followed by titration. A pH meter or indicator is used. Recall that indicators are dyes that change color with changes in the pH of a solution. Consider the reaction of a carbonate or hydrogen carbonate with HCl and write balanced equations. Consider the accuracy and precision of a titration. How many trials will be needed? Do you expect to see a difference between the titration of a carbonate and a hydrogen carbonate? If you are not sure, titrate the known carbonate and the known hydrogen carbonate then compare the results before titrating the unknown.

Safety

Eye protection must always be worn in the laboratory. The acid and base used in this experiment may irritate the skin. Wash any spills immediately with water. Reactions can be quite vigorous; use caution when mixing reagents.

Experimental

If you have an estimate of the molar mass of your unknown, you can use this value to estimate the amount of unknown to use. If you are using the known carbonate or hydrogen carbonate, you know the molar mass and can calculate the amount to titrate. A titration should use 20 to 30 mL of titrant. In this case, the titrant is 0.1 M HCl. To measure the unknown, weigh by difference to the nearest 0.1 mg an amount of the unknown into the Erlenmeyer flask. Dissolve with water. This may take some trial and error.

For indicator titrations, start with phenolphthalein and titrate to the color change. Record the volume of acid used. Add methyl orange/indigo carmine <u>or</u> methyl red indicator. Methyl orange/indigo carmine changes color from green to purple at a pH of about 4 while methyl red changes from yellow to red. The methyl red end point is at the orange color (red + yellow). Record the volume of acid used. Estimate the molar mass of the unknown. Estimate the accuracy and precision. A known sample of carbonate or hydrogen carbonate can be titrated for comparison.

For pH titrations, prepare the pH meter according to your instructor's directions. Add the acid in small increments of about 0.5 to 1 mL until the pH begins to change rapidly and then decrease the amount of each increment to 0.5 mL or less. Plot the pH vs. volume of acid added. Determine the equivalence point(s) and the molar mass of the unknown. See **Finding the Equivalence Point on a pH Titration Curve** in this manual. Estimate the accuracy and precision.

Disposal Methods

Solutions may be flushed down the drain unless otherwise directed by your instructor. Keep unused unknown in a desiccator.

Questions

1. What is the molar mass of the unknown?
2. What are the precision and accuracy of the data?

Eudiometry[1]

Eudiometry is used to determine the moles of unknown in a weighed sample by measuring the volume of gas evolved from the unknown. The reaction of a carbonate or hydrogen carbonate with acid results in the evolution of gas. In the mass loss experiment, the mass of the gas evolved is found. In this experiment, the volume of the gas is found. The gas is collected in a buret over a column of water. See Figure 1. By proper application of Dalton's law of partial pressures and the ideal gas law, the moles of unknown are found. First, a correction is applied for the pressure of water vapor over the column of water. Knowing the temperature, pressure, and the volume of gas, calculate the moles of gas. From the known mass of the unknown and the moles of gas evolved, the molar mass of the unknown can be calculated.

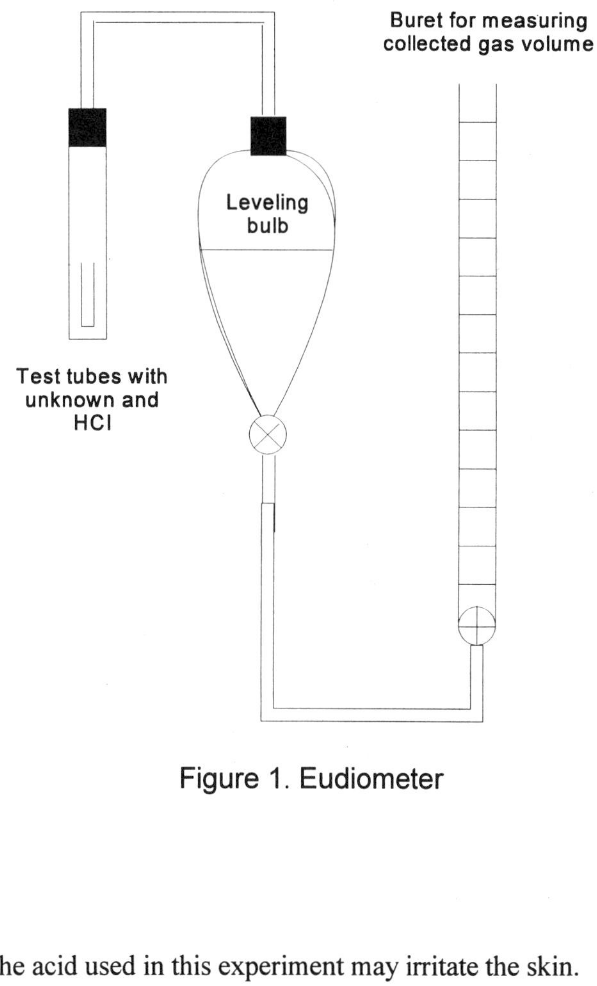

Figure 1. Eudiometer

Safety

Eye protection must always be worn in the laboratory. The acid used in this experiment may irritate the skin. Wash any spills immediately with water. Reactions can be quite vigorous, so use caution when mixing reagents.

Experimental

Weigh by difference approximately 0.15 g of unknown to the nearest 0.1 mg in a test tube. Add approximately 2 mL of 6 M HCl to a second test tube. Insert the smaller test tube into the larger as in Figure 1 without mixing the reagents. Connect the test tubes to the eudiometer and record the initial volume of water in the buret by bringing the leveling bulb to the same height as the water. Mix the unknown and the acid taking care to not get the acid into the rubber stopper. Once the column of water stops rising, equilibrate the pressure in the system by bringing the leveling bulb to the same level as the water in the buret. Read the final volume. Consider what data you will need to apply the gas laws and calculate the number of moles of gas evolved. Carry out a sample calculation to be sure you are recording the proper data. If a known is used, estimate the accuracy of the method.

Disposal Methods

Solutions may be flushed down the drain unless otherwise directed by your instructor. Keep unused unknown in a desiccator.

Questions

1. What is the reaction of the unknown with HCl? Write the balanced equation.
2. What is the gas you are measuring?
3. How will the solubility of the gas in water and acid affect the moles of gas measured? What steps can be taken to reduce this error?
4. What is the effect of the pressure of water vapor on the measurement?
5. Estimate the accuracy and precision of this experiment.

Physical Properties

Safety

Eye protection must always be worn in the laboratory. Wash any spills immediately with water. Some reactions can be quite vigorous so use caution when mixing reagents.

Experimental

Describe the unknown. Consider color, texture and any other physical characteristics you observe. Determine the solubility in various solvents and compare these with the list in your **Preparation**. Recall that solubility is given in g/100 mL. You will not need to use 100 mL but a much smaller volume.

Disposal Methods

Place any solutions in the appropriately labeled bottles or beakers. Keep unused unknown in the desiccator.

Question

1. Do the physical properties help you eliminate any compounds?

Flame Tests

When heated in a Bunsen burner flame the cations of Groups 1 and 2 exhibit characteristic colors. Consult your textbook or the CRC for information concerning the colors of these metals. Several salts of Group 1 and 2 metals are available for comparison.

Safety

Safety glasses must be worn at all times in the laboratory. CAUTION: POTENTIALLY DANGEROUS UV LIGHT MAY BE EMITTED. DO NOT LOOK AT THE FLAME LONGER THAN NECESSARY TO OBSERVE THE COLOR.

Experimental

Obtain a small amount of the unknown on the end of a scoopula. Hold the scoopula in the hot portion of the flame and observe the color of the flame. NOTE: Do not drop the solid into the Bunsen burner.

Disposal Methods

Wash the scoopula with plenty of water.

Question

1. What is the color associated with each Group 1 and 2 cation?

General questions to be answered considering all of the data obtained by your group

1. Which experiments had the most precise results?
2. Which experiments had the most accurate results?
3. What is the identity of your unknown based on your experimental results?

References

1. "A Carbonate Project Introducing Students to the Chemistry Lab" by Emily Dudek, *Journal of Chemical Education*, **1991**,68, 948-950.

2. "A Simple Quantitative Synthesis: Sodium Chloride from Sodium Carbonate" by Marvin Gold, *Journal of Chemical Education*, **1988**, 66, 731.

Preparation
Carbonates and Hydrogen Carbonates

Name ______________
Section ______________
Date ______________

Read

1. This experiment.
2. About carbonates in your textbook.

Questions

1. Look up and record in your notebook the solubilities of the Group 1 and 2 carbonates and hydrogen carbonates (g/100 mL in water; list other possible solvents). Suggested references are the CRC Handbook of Chemistry and Physics, and Lange's Handbook of Chemistry

2. What is the formula of a metal carbonate? ____________

3. What is the formula of a metal hydrogen carbonate? ____________

4. Write a balanced equation for the reaction of:
   - a carbonate with HCl
   - a hydrogen carbonate with HCl
   - a carbonate with NaOH
   - a hydrogen carbonate with NaOH

Practice Problems

1. The following reaction occurs in aqueous solution. Write the balanced molecular equation and the net ionic equation including the phase labels.

$$CuCO_3(s) + H_2SO_4(aq) \rightarrow$$

2. A 2.56 g sample of a compound of cobalt and chloride was dissolved in water. The Cl^- was precipitated with excess silver nitrate as AgCl. Write the net ionic equation for the precipitation reaction. The mass of the precipitate was 4.14 g. How many grams of Cl^- were in the original cobalt sample? What is the percent by weight of Cl^- in the cobalt compound?

3. When warmed with sulfuric acid, formic acid decomposes according to the equation:

$$HCO_2H(l) \rightarrow H_2O(l) + CO(g)$$

A total of 4.63 L of CO was collected over water at 23 °C and 725 mmHg. What mass of HCO_2H was consumed?

4. Assume that you react exactly 1.00 g of Na_2CO_3 with HCl. How many grams of NaCl are produced?

Starting with 1.00 g of $NaHCO_3$, how many grams of NaCl are produced?

Explain how the difference in the mass of the products can be used to help distinguish a carbonate from a hydrogen carbonate.

Results
Beverages

Name ____________________
Section ____________________
Coworkers ____________________
____
____
____
Date ____________________

Experiment:

Record data and observations in an appropriate format.

Carbonates and Hydrogen Carbonates-- Report Guidelines

- A First Draft can be evaluated ahead of the report due date by appointment with the instructor.
- Be sure all calculations are correct before the report is turned in.
- All members of a group should contribute to the report and edit drafts and the final version. Each report will receive a single grade and all group members will receive that grade.
- The report must be double spaced.

The following guidelines should be followed when writing scientific reports:

I. Title - descriptive of the project

II. Introduction

A. Broad statement of purpose of project

B. Types of experiments performed, but not experimental or procedural details

C. Include theory and information gathered along with background information

III. Experimental

A. Specific details of experimental procedures to allow a reader to reproduce them

B. Information which is general knowledge <u>should not</u> be included, e.g. do not say "take initial and final reading of buret", but rather "titration volumes were as follows".

C. Group experiments by category, writing a general procedure for all similar experiments; then include data for results of specific tests in each category. For example, in determining the molar mass of the unknown by simple weight loss, write the procedure once. Include one sample calculation for determining the molar mass and summarize the important data in a table. Include the data necessary to carry out the calculations as well as the results. Give an estimate of the accuracy and precision of the experiment. For pH titrations, a graph which can be read to the precision of the measurements is sufficient. DO NOT include tables of pH - volume data.

IV. Discussion

A. Identify your unknown, the significance of the experimental results, and in particular, the logic applied to identify the unknown.

B. Summary of any results not included in the tables above

C. Proposals for further research

V. Conclusion

Brief summary of what was learned

VI. References

A. Numbered list of all references used

B. Reference within report to particular sources when appropriate

VII. Other Considerations

grammar
sentence structure
spelling
format
First person passive voice should be used for the experimental
Use consistent voice in verb usage.

VIII. Tables and graphs should be consecutively numbered with a descriptive title and appropriate footnotes.

PROJECT EVALUATION FORM
CARBONATES AND HYDROGEN CARBONATES

NAME:_______________ DATE:_____

Were you project director? ____ YES ____ NO

1. List the experiments that you did:

2. Using a number from 1 to 10, what grade would you give yourself? Justify your answer.

3. Using a number from 1 to 10, what grade would you give each of your partners. Give examples from lab, outside meetings, and contributions to the report to justify your answers. Please use full names.

4. What did the lab instructor do that was helpful/unhelpful in completing your project?

Halogens

2 lab periods; work in groups of 2 or 3. Complete the Preparation page each week.

Goals

- To explore the solubility, oxidizing and reducing strength of selected Group 17 (VIIA) elements and ions
- To design and use qualitative identification techniques
- To apply the techniques of gravimetric analysis
- To prepare a scientific report, as a group, detailing the results of these experiments

Background

Many properties observed for a family of compounds are related to periodic trends. For example, atomic and ionic size, electron configuration, electronegativity or electron affinity vary with placement in the periodic table and are expected to influence chemical reactivity. The experiments in this project may be used to study this assertion. Keep in mind that each experiment is an integral part of the project and although it may stand alone, you are looking for trends in the data to ascertain the influence of periodic properties on the chemical reactivity. Also, this is an experiment so not every aspect may work correctly the first time. Use the knowledge gained from initial trials to refine your procedure and improve the results.

For the best results with the solubility experiments, you will need two weeks so you should plan accordingly. Moreover, when developing your plan, consider how many trials will be necessary to give you confidence in your results.

Solubility is defined as the maximum amount of solute that will dissolve in a given amount of solvent at a specified temperature and pressure, i.e. the amount of solute necessary to saturate a solution at room temperature and pressure. Generally, the solubility is reported in grams of solute per 100 mL of solvent as in the CRC Handbook of Chemistry and Physics or Lange's Handbook of Chemistry. Although the solubility is reported per 100 mL of solution, you will not have to prepare 100 mL of solution. In fact, 1.000 mL samples of the solution to be tested are used. Using this knowledge and the number of trials needed, decide what volume of saturated potassium halide solution must be prepared. In your proposal, describe how you would prepare a saturated solution. Is there a way to estimate how much of the potassium halide would saturate the volume of solution you will need?

Once you have prepared the saturated solutions to be tested, you will apply gravimetric analysis techniques to determine the amount of halide in the original saturated solution. Gravimetric analysis experiments use the mass of a product of known molar ratio to calculate the amount of a species in an unknown solid or solution. In this case, the mass of the silver halide is used to calculate the amount of potassium halide in the saturated solution. Therefore, the solid formed for each trial must be separately isolated, dried and the mass found. Gooch crucibles fitted with filter paper are designed for this purpose. They are ceramic so that they are stable when heated.

A qualitative identification scheme is a way of tracking reactions. Without measuring exact amounts of reactants, one can determine if a reaction has occurred by looking for evidence of the reaction. For example, if two solutions form a precipitate upon mixing, we conclude a reaction has occurred. Similarly, if the color of a solution changes upon mixing two reagents we conclude that a reaction has occurred.

To explore the oxidizing strengths of the halogens, Cl_2, Br_2, and I_2, in terms of the oxidizing and reducing strengths of the corresponding halides, Cl^-, Br^-, and I^-, a qualitative analysis scheme is used to determine when a reaction has occurred. Physical signs that a reaction has occurred include evolution of heat, evolution of gases, formation of a precipitate, color changes, and dissolution of a solid. A qualitative analysis scheme can be developed based on the difference in solubility of the halogens in water and in mineral oil. By observing the color change of the mineral oil when mixed with aqueous solutions of the halogens and solutions of sodium halides a qualitative analysis scheme is developed. The presence or absence of the halogens is based on the colors observed in each solvent after mixing.[1] The oxidizing strength of the halogens can be determined by reacting each aqueous halogen solution with the sodium halide solution of the other halides. In addition, each halogen solution is reacted with elemental Cu. The halide strength as reducing agents can be determined by reaction with the other halogens and also with solutions of Cu^{2+} and $KMnO_4$. Group members should divide the work in this project such that all members have the opportunity to observe and compare results of the qualitative analysis. Keep in mind that the solubility experiment requires two weeks.

Is there a periodic trend in the solubility and oxidizing and reducing strengths of the elements in Group 17?
What properties account for any trends observed?

Equipment and Materials

test tubes 13 mm
Gooch crucibles and filters
drying oven
desiccator
Parafilm
mineral oil
1 M aqueous solutions of NaCl, NaBr, NaI
1 M $CuSO_4$
0.01 M $KMnO_4$ in 1 M H_2SO_4
0.3 M $AgNO_3$
aqueous solutions of Cl_2, Br_2, I_2
Cu turnings
solid KCl, KBr, KI

Safety

Eye protection must always be worn in the laboratory. Silver nitrate stains the skin brown. The permanganate is in a strong acid solution and can burn the skin. Wash immediately if contact with skin occurs.

Solubility of Potassium Halide Salts

Experimental

Record results in your notebook or on the **Results** page, as directed. This is a 2 week experiment and must be started during the first week. Prepare a saturated solution of each halide salt: KCl, KBr, and KI. When preparing the saturated solution, consider the total volume of solution needed so you can easily pipet each of the 1.000 mL samples, but not have an excess remaining. Be sure the solutions are at room temperature before removing the sample to precipitate.

Find the maximum solubility of each potassium halide by precipitation of 1.000 mL samples of the halide with 0.3 M $AgNO_3$, and gravimetric analysis of the product. How do you know the precipitation is complete? Once the precipitation is complete, add 5 mL of 0.3 M $AgNO_3$. The silver halide salts formed are light sensitive; avoid prolonged exposure to room light. For best results, digest the precipitate by gently boiling the solution for 5-10 minutes, cool to room temperature, then further cool in an ice bath before filtering. Alternately, store in the dark until week 2. Isolate the solid by filtering through a dried Gooch crucible at constant mass. Rinse the beaker with 5 mL of 0.02 M HNO_3 and wash the precipitate with an additional 5 mL of the acid. Dry the solid to constant mass.

Disposal Methods

Combine filtrates and place in the bottle or beaker of silver waste. Place precipitates in the bottle or beaker of silver halide waste.

Qualitative Identification Scheme and Redox Reactions

Experimental

Examine and note the colors of the halogens and their halides in aqueous solution. Next, mix several drops of mineral oil with each aqueous solution, shake vigorously and describe your observations. Do the solvents mix? Why or why not? Which solvent is more dense? What can you conclude about the solubility of the halogens and halides in mineral oil? Use this information to devise a method to identify the presence of a halogen in a mixture based on the colors you have observed.

Oxidizing Agents

Experimental

Find the relative strengths of the halogens as oxidizing agents by exploring their reactions with the other halides and then with Cu turnings. Add approximately equal volumes of the aqueous halogen and the halide solution to a 13 mm test tube. Mix the solutions by vigorously shaking the test tube. The top may be covered with Parafilm. Note any changes in the solution. Use the qualitative identification scheme proposed above and several drops of mineral oil to help you determine when a reaction has occurred and what the products are.

Add a small piece of Cu turning (approximately 1 to 2 cm long) to test tubes with about 5 mL of aqueous halogen. Cover with Parafilm and allow to sit for at least one hour before testing for reaction with mineral oil. What can you conclude about the halogens as oxidizing agents?

Reducing Agents

Experimental

Determine the relative strengths of the halides as reducing agents by exploring the reactions of each with $CuSO_4$ and $KMnO_4$. Mix approximately equal volumes of each solution in 13 mm test tubes. Use the qualitative scheme developed above to help you decide if a reaction has occurred and what products may be present. Also consider the reactions of halogens and halides above. What can you conclude about the halides as reducing agents?

Disposal Methods

All halogen wastes are collected in the appropriately labeled waste bottles or beakers. Do not wash down the drain.

Questions

1. Solubility of Potassium Halides

1. Is there a trend in solubility for the potassium halides?
2. Can the results be explained by periodic properties?
3. What precision and accuracy did you obtain for this experiment? What quantities need to be measured precisely? Why?
4. What is the precipitate collected? Write a balanced chemical equation for each reaction.
5. Calculate the solubility of potassium halide per 100 mL of solution with the correct number of significant figures. Estimate the precision of your experiment. Compare your values to those found in the CRC Handbook of Chemistry and Physics or Lange's Handbook of Chemistry.
6. Arrange the potassium halides in order of solubility.

2. Qualitative Analysis Scheme

1. Mineral oil is a nonpolar solvent. Is water polar or nonpolar?
2. Are the halogens polar or nonpolar? Are the halides polar or nonpolar?
3. Explain how the different solvents were useful in the qualitative analysis scheme. Note that extraction is occurring in some of these reactions.

3. Oxidizing and reducing properties

Prepare a table showing reactions attempted. Include reactants, observations, and conclusions about whether a reaction occurred and what products were formed.

1. Write balanced chemical equations for reactions that occurred.
2. Arrange the halogens in order of their strength as oxidizing agents.
3. Arrange the halides in order of their strength as reducing agents.

4. Conclusions

Explain these trends based on the position of the halogens/halides in the periodic table, using atomic and ionic size, structure, electronic configuration, electron affinity, and other chemical principles as aids.

Reference

"The Suitability of Mineral Oil as a Halogen Extraction solvent" by Thomas M. Scott, *Journal of Chemical Education*, 1991, 68, 950.

Preparation
Halogens - Week 1

Name ______________
Section ______________
Date ______________

Read

1. This experiment.
2. About halogens in your text.
3. About solubility and solvation energy, electronegativity and electron affinity, and the periodic table in your text.

Within your group, devise an experimental plan and decide who in the group will be doing which experiments, keeping in mind that the solubility experiments require two weeks and that the qualitative analysis is best understood by observing all of the reactions. Prepare an introduction in your notebook describing the goals of your experimental plan, and the particular experiments you will be doing.

Questions

Define the following terms:

- oxidizing agent
- reducing agent
- saturated solution
- polar molecule
- nonpolar molecule
- extraction
- electron affinity
- electronegativity
- lattice energy
- solvation energy

Practice problem

A 75.0 mL solution of $BaCl_2$ was analyzed for the Ba content. The Ba^{2+} was precipitated with an excess of sodium sulfate. The $BaSO_4$ collected had a mass of 1.245 g. What was the molarity of the original solution? How many grams of AgCl would be collected if the precipitation was carried out with silver nitrate?

Preparation
Halogens - Week 2

Name ____________

Section ____________

Date ____________

Problems

1. Balance the following reactions in acid or base as indicated:

$MnO_4^- + I^- \rightarrow Mn^{2+} + I_2$	(acid)
$Cl_2 \rightarrow Cl^- + ClO_3^-$	(base)
$Zn + NO_3^- \rightarrow NH_3 + Zn(OH)_4^{2-}$	(base)
$Cr_2O_7^{2-} + I^- \rightarrow Cr^{3+} + IO_3^-$	(acid)

2. Prepare a table that includes the following information for elements in Group 17 (VIIA):

name of element
symbol
formula for most common form of pure element
most common ion
name of most common ion
electronic configuration of monatomic element
electronic configuration of most common ion

Results
Halogens - Week One

Name ____________
Coworker(s) ____________
____

Section ____________
Date ____________

Solubility of potassium halides

Reaction of KX and silver nitrate:

Table 1. Data for the solubility of __________

Trial no.	g used of KX	volume sat'd solution	volume $AgNO_3$	g silver halide	solubility g/100 mL

Observations

Results
Halogens - Week Two

Name ______________
Coworker(s) ______________
____
Section ______________
Date ______________

Qualitative Analysis

Table 2. Color of solutions and reactions with mineral oil

Solution	Color	Mineral Oil

Table 3. Reactions upon Mixing Solutions

Solution 1	Solution 2	Result of Mixing	Reaction	Balanced Equation

Periodic Trend in Solubility: ______________________________

Ordering of Oxidizing Strength: ______________________________

Ordering of Reducing Strength: ______________________________

Answers to Questions

Halogens-- Report Guidelines

- A first draft can be evaluated ahead of the report due date by appointment with the instructor.
- Be sure all calculations are correct before the report is turned in.
- All members of a group should contribute to the report and edit drafts and the final version. Each report will receive a single grade and all group members will receive that grade.
- The report must be double spaced.

The following guidelines should be followed when writing scientific reports:

I. Title - descriptive of the project

II. Introduction

A. Broad statement of purpose of experiment

B. Types of experiments performed, but not experimental or procedural details

C. Include theory and information to be gathered along with background information

III. Experimental

A. Specific details of experimental procedures to allow a reader to reproduce them

B. Information which is general knowledge should not be included, e.g. do not say "take initial and final reading of buret", but rather "titration volumes were as follows"

C. Group experiments by category, writing a general procedure for all similar experiments; then include data for results of specific tests in each category. For example, in determining solubility of halide salts, write a general procedure for preparing the saturated solutions and precipitating the silver halide. Then present a table giving specific amounts of silver salt recovered for each halide, and the calculated solubility of each salt. If appropriate, include a sample calculation and put results in a table.

D. Present relevant observations and data in tables and graphs if appropriate.

IV. Discussion

A. Significance of experimental results in terms of periodic properties. Did you demonstrate any trends in periodic properties based on your results? Were they the trends you expected or would have predicted? Why or why not?

B. Summary of results presented in the tables in part III, but not a wholesale repetition of the above. Refer to the data in the tables.

C. Proposals for further research

V. Conclusions

VI. References

A. Numbered list of all references used

B. Reference within report to particular sources when appropriate

VII. Other Considerations

grammar
sentence structure
spelling
format
past tense, passive voice for experimental
consistent voice in verb usage
concise writing, not repetitive

VIII. Tables and graphs should be consecutively numbered with an descriptive title and appropriate footnotes.

PROJECT EVALUATION FORM
HALOGENS

NAME:________________ DATE:______

Were you project director? ____ YES ____ NO

1. List the experiments that <u>you</u> did:

2. Using a number from 1 to 10, what grade would you give yourself? Justify your answer.

3. Using a number from 1 to 10, what grade would you give each of your partners. Please include first and last names for each. Give examples from lab, outside meetings, and contributions to the report to justify your answers.

4. What did the lab instructor do that was helpful/unhelpful in completing your project?